SCIENCE FOUNDATIONS

Cell Theory

SCIENCE FOUNDATIONS

The Big Bang
Cell Theory
Electricity and Magnetism
Evolution
The Expanding Universe
The Genetic Code
Germ Theory
Gravity
Heredity
Kingdoms of Life
Light and Sound
Matter and Energy
Natural Selection
Planetary Motion
Plate Tectonics
Quantum Theory
Radioactivity
Vaccines

SCIENCE FOUNDATIONS

Cell Theory

ALLAN B. COBB

Science Foundations: Cell Theory

Copyright © 2011 by Infobase Learning

All rights reserved. No part of this book may be reproduced or utilized in any form or by any means, electronic or mechanical, including photocopying, recording, or by any information storage or retrieval systems, without permission in writing from the publisher. For information, contact:

Chelsea House
An imprint of Infobase Learning
132 West 31st Street
New York, NY 10001

Library of Congress Cataloging-in-Publication Data
Cobb, Allan B.
  Cell theory / Allan B. Cobb.
    p. cm. — (Science foundations)
  Includes bibliographical references and index.
  ISBN 978-1-60413-058-4 (hardcover)
 1. Cells—Popular works. 2. Cytology—Popular works. I. Title. II. Series.

QH582.4.C63 2011
571.6—dc22   2010026884

Chelsea House books are available at special discounts when purchased in bulk quantities for businesses, associations, institutions, or sales promotions. Please call our Special Sales Department in New York at (212) 967-8800 or (800) 322-8755.

You can find Chelsea House on the World Wide Web at
http://www.infobaselearning.com

Text design by Kerry Casey
Cover design by Alicia Post
Composition by EJB Publishing Services
Cover printed by Yurchak Printing, Landisville, Pa.
Book printed and bound by Yurchak Printing, Landisville, Pa.
Date printed: June 2011
Printed in the United States of America

10 9 8 7 6 5 4 3 2 1

This book is printed on acid-free paper.

All links and Web addresses were checked and verified to be correct at the time of publication. Because of the dynamic nature of the Web, some addresses and links may have changed since publication and may no longer be valid.

Contents

1	Introduction to the Cell Theory	7
2	Tools for Studying Cells	22
3	Developing the Cell Theory	34
4	The Neuron Theory	47
5	DNA and the Cell Theory	55
6	The Life of a Cell	67
7	Applying and Questioning the Cell Theory	78
8	The Cell Theory and Modern Biology	88
	Glossary	94
	Bibliography	96
	Further Resources	97
	Picture Credits	98
	Index	99
	About the Author	104

Introduction to the Cell Theory

The cell theory states that all organisms are composed of similar units of organization that are called **cells**. While this seems obvious today, this concept was a giant conceptual leap in biology. While Robert Hooke discovered cells in 1665, from there, it took scientists more than 150 years to develop cell theory. The theory was formally articulated in 1839 by Theodor Schwann, Matthias Jakob Schleiden, and Rudolf Virchow and has remained as the foundation of modern biology. The idea predates other great paradigms of biology including Darwin's theory of evolution, which was first formulated in 1859; Mendel's laws of inheritance, formulated in 1865; and the establishment of comparative biochemistry in 1940. Today, the cell theory is one of the overarching concepts in biology.

CLASSICAL CELL THEORY

The classical cell theory is composed of three basic statements. Together, these three statements make up the classical cell theory. When researchers Theodor Schwann (1810–1882), Matthias Jakob Schleiden (1804–1881), and Rudolf Ludwig Karl Virchow (1821–1902) proposed the classical cell theory in 1839, they based this on their work and observations on cells. Schwann and Schleiden suggested that cells were the basic unit of life. Their theory accepted the first two tenets of classical cell theory. However, the cell theory of Schwann and Schleiden differed from classical cell theory in that it proposed a method of spontaneous crystallization called free cell formation. In 1858, Virchow concluded that all cells come from pre-existing cells, thus completing the classical cell theory. The story of how they came up with this theory will be covered in a later chapter.

All organisms are made up of one or more cells. This means that all plants and all animals are made up of cells. Single-celled organisms such as protists and **bacteria** are made up of a single cell. Multicellular organisms have more than one cell. Whether single-celled or multicelled, each cell carries out a variety of functions. In single-celled organisms, the cell carries out all functions that keep the organism alive. The cell takes in food, breaks it down into energy, uses the energy to grow, and uses energy to reproduce. In multi-celled organisms, each cell performs these same functions, but they also work with other cells in the organism. In multicellular organisms, cells become specialized and only carry out a limited number of functions. Some cells conduct nerve signals, some cells produce chemical compounds needed by other cells, and some cells form organs. Exactly how cells function and what they need to function will be covered later.

Cells form the basic unit of life and structure. All organisms—plant or animal—are made up of cells. The cell is the smallest unit of life and the basic structure. An organism simply cannot be alive if it does not have one or more cells. Cells are remarkably similar among all living organisms. All cells basically operate in the same

Classical Cell Theory

The following points provide the basis for the classical cell theory:

- All organisms are made up of one or more cells.
- Cells are the basic unit of life and structure.
- All cells come from pre-existing cells.

manner, and they have similar sets of structures within them. All of this makes cells the building block of life.

All cells come from pre-existing cells. One function of cells is to reproduce. Cells make more cells to replace themselves and to allow multicelled organisms to grow. Cells do not simply appear and start multiplying. Parents give cells to their offspring that develop and grow. Scientists widely accept this tenet of the cell theory. However, this tenet also presents us with a mystery: where did the first cell come from? If all cells come from pre-existing cells, how did the first cell form? This question will be explored in Chapter 7.

MODERN CELL THEORY

Over the past 170 years, scientists have learned more and more about cells. The basic tenets of the classical cell theory all still hold true today. In fact, these are the ones listed in textbooks as the definition of the cell theory. However, as scientists have learned more, they have added to the classical cell theory and, in some cases, refined some of the wording. While the classical cell theory is still true, the expanded modern cell theory is more accurate. The modern cell theory adds tenets about the passing of **genetic material** from parents to offspring, the chemical composition of cells, and energy flow within cells. These concepts were unknown at the time that the classical cell theory was established in 1839.

The cell theory is one of the overarching themes of biology. It serves as a cornerstone upon which scientists study cells, organisms, and life. Almost all work in biology somehow ties back into the cell theory. The cell theory is well researched, well accepted, and universally agreed upon by scientists.

However, with all the work that has been done on the cell theory, one might wonder why it is both called a theory and considered a law.

WHAT IS SCIENTIFIC THEORY?

To nonscientists, a theory sounds like a nebulous thing, as though scientists are not sure about what they believe to be fact. However, this is not even close to reality. To a scientist, a theory is an explanation or model based on observation, experimentation, and reasoning, especially one that has been tested and confirmed as a general principle that helps to explain and predict natural phenomena. A scientific theory must be based on a careful and rational examination of the facts. A clear distinction needs to be made between *facts* (things that can be observed and/or measured) and *theories* (explanations that correlate and interpret the facts).

Theories may be good, bad, or indifferent. Theories may be well established by the factual evidence, or they may lack credibility. Before a theory is given any credence in the scientific community, it must be well tested and subjected to peer review. This means that the proposed theory must be published in a legitimate scientific journal in order to provide the opportunity for other scientists to evaluate the relevant factual information and publish their conclusions. Scientists continue to test theories and apply them to new situations. Over time, the theory may remain the same, become modified in light of new research, or it may be proven wrong and dismissed. In science, theories are used to describe phenomena in a uniform way. If a theory is proved beyond all doubt, it may become a scientific law.

Modern Cell Theory

The following points provide the basis for the modern cell theory:

- All known living things are made up of cells.
- The cell is the structural and functional unit of all living things.
- All cells come from pre-existing cells by division. (Spontaneous generation does not occur.)
- Cells contain hereditary information that is passed from cell to cell during cell division.
- All cells are basically the same in chemical composition.
- All energy flow (metabolism and biochemistry) of life occurs within cells.

It is actually difficult for a scientific theory to become a scientific law because there is so much to learn about any subject.

The cell theory has stood for almost 200 years. During that time, it has had minor changes made to it and it has had tenets added to it. In the future, some of the current tenets may be dropped or expanded on. New tenets may be added if research dictates the need.

CHEMICAL COMPOSITION OF CELLS

Cells may be the smallest unit of an organism, but cells are made up of atoms. The atoms in cells are bonded together to make many different chemical compounds. The most common compound in cells is water. Water is found in all cells and is considered essential to life as we know it. Water is important in cells because water makes it possible for food molecules to move throughout the cell to different locations. Water is also important because it carries away wastes. Water is an important solvent that dissolves salts and creates ions that are important to various cellular processes.

Cells are also composed of different organic molecules. These organic molecules are carbohydrates, lipids, proteins, and nucleic acids. Organic molecules contain carbon. Carbon is one of the more common elements that are found in cells. In fact, all life that we know of is called carbon-based life. Every cell is made up of carbon-based organic molecules.

Carbohydrates are organic molecules made up of carbon, hydrogen, and oxygen. These atoms occur in a ratio of 1:2:1. The general chemical formula for carbohydrates is $C_n(H_2O)_n$, where n is the number of carbon atoms and water. For example, glucose is a common sugar found in cells. Glucose molecules all contain 6 carbon atoms, so n is equal to 6. The chemical formula for glucose is $C_6H_{12}O_6$.

The carbohydrate glucose plays an important role because cells use it to store and transport energy. Glucose is made in plants by combining carbon dioxide and water in a process known as photosynthesis. Simple carbohydrates, such as glucose, may be linked together into chains to make polysaccharides. Polysaccharides are one way that organisms store glucose for future use. Plants make several kinds of glucose-storage polysaccharide, such as starch and cellulose, while animals make one called glycogen. These polysaccharides may be broken down into glucose when energy is needed at a later time.

Lipids are another type of organic molecule found in cells, but they do not dissolve in water. Commonly called fats and oils, lipids also include larger molecules, such as waxes, and have many different functions in organisms. Lipids may be used as energy storage, parts of cell membranes, and in various metabolic reactions.

An important function of lipids in cells is in the formation of cellular membranes. All cells are surrounded by membranes and membranes usually surround the different parts of cells. These membranes are made of lipids with attached phosphate molecules. These molecules are called phospholipids. Phospholipids are important because the phosphate "head" has a slight ionic charge, so it attracts water molecules while the "tail" repels water. These properties make membranes that separate cells from the outside environment.

Proteins are another important group of organic molecules found in cells. The cells that make support structures make use of proteins. Other proteins are important for the metabolism of all living cells. Some proteins, called **enzymes**, are important catalysts for biochemical reactions that take place in cells. Enzymes are important because they allow certain chemical reactions to take place within cells that would be impossible without them. This allows cells to maintain an internal balance through the control of chemical reactions. The ability to maintain their internal environment is vital for cells in order for them to survive in many different environmental conditions.

Proteins are long chains of amino acids. Cells assemble amino acids into specific proteins. The sequence in which amino acids are assembled makes a protein into a unique and specific shape. This shape helps determine the function of the protein. In humans, all the proteins are assembled from just 20 amino acids. Some proteins may be composed of hundreds of amino acids linked together in a specific order. Cells obtain the needed amino acids to build proteins from their food.

Nucleic acids are the final important group of organic molecules that are found in all living organisms. Nucleic acids are made up of long chains of subunits called nucleotides. A nucleotide consists of a five-carbon sugar, a phosphate group, and an organic base. The carbon atoms in a nucleotide are shaped like rings with the phosphate attached to one side and the organic base to the other. The main function of nucleic acids is to store hereditary information that is used to build proteins. Most cells use a nucleic acid called **deoxyribonucleic acid**, or **DNA**, to fulfill these functions. The other nucleic acid is **ribonucleic acid**, or **RNA**. RNA has a slightly different structure than DNA, and it performs a variety of roles in the process of making proteins.

These classes of organic chemicals are found in all cells. Single-celled organisms and multicelled organisms use these same organic molecules for the same functions. There are some differences in specific proteins, but they are all very similar.

All cells use energy to grow, move, and process information. This energy comes from food. Plants make their own food from sunlight while animals must eat plants or other animals to get energy. Even single-celled organisms either get food by making it or eating it. To either make food or break down other organisms into food requires a series of chemical reactions to store or release energy. The sum of all these chemical reactions is called metabolism.

A typical cell is made up of an outer **plasma membrane**, a **nucleus**, **cytoplasm**, and **organelles**. Some cells have more complex interior structures than others, but all cells grow and reproduce. The nucleus is surrounded by a membrane to separate it from the rest of the interior of the cell. It is also the control center of the cell and contains the genetic information for the cell. The cytoplasm is made up of all the material located between the outer plasma membrane and the nuclear membrane. All cells depend on their organelles to carry out specific functions within the cell. Finally, all cells are surrounded by a plasma membrane that separates the interior of the cell from the external environment.

The plasma membrane of a cell is made up of two layers of phospholipid molecules. The two layers are oriented so that the lipid ends of the molecule meet in the center of the two layers. This arrangement places the phosphate end, with its slight electric charge, toward both the inner and outer surface of the membrane. Because of the arrangement of the phospholipids, water cannot pass through the membrane. The plasma membrane is very important because it separates the interior of the cell from the external environment.

The plasma membrane itself does not allow water or other molecules to pass between the cell and the environment. However, the

14 CELL THEORY

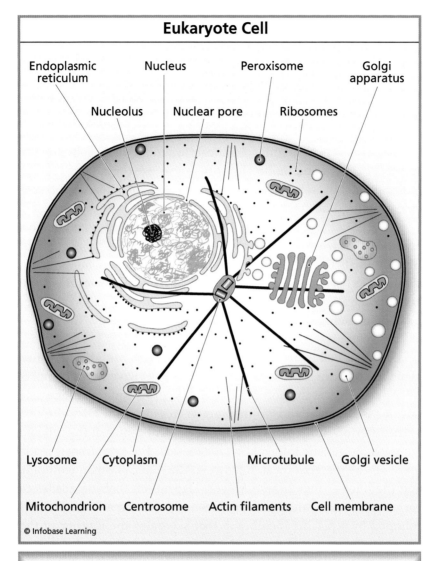

Figure 1.1 Animals have eukaryote cells. In these cells, the nucleus contains the DNA genome and an assembly plant for ribosomal subunits (the nucleolus). The endoplasmic reticulum (ER) and the Golgi apparatus work together to modify proteins, most of which are destined for the cell membrane. These proteins are sent to the membrane in Golgi vesicles. Mitochondria provide the cell with energy in the form of ATP. Ribosomes—some of which are attached to the ER—synthesize proteins. Lysosomes and peroxisomes recycle cellular material and molecules. The microtubules and centrosome form the spindle apparatus for moving chromosomes to the daughter cells during cell division.

cell needs to get water, minerals, and nutrients from the outside environment into the cell, and it needs to move wastes from inside the cell to the outside. Different types of protein molecules embedded in the plasma membrane perform these tasks. Some of these protein molecules penetrate all the way through the plasma membrane while others only penetrate partway from either the inside or the outside of the plasma membrane. These proteins are often attached to different carbohydrate molecules that aid in passing molecules into and out cells. Some of the important functions of these protein pathways involve the transport of inorganic ions, water molecules, and dissolved gases.

The nucleus of a cell is spherical in shape and houses the DNA. The nucleus is separated from the rest of the cell by a nuclear membrane. The DNA is the hereditary material that provides instructions to the cell for making all proteins needed by the cell. The DNA is typically tightly coiled and constitutes the core of a structure known as the **chromosome**. The nucleus contains another structure called the nucleolus. A typical cell has two nucleoli. The nucleolus is small and spherical and contains RNA that is used in the production of a specialized organelle called ribosomes.

All the material in a cell between the plasma membrane and the nuclear membrane is called the cytoplasm. The cytoplasm is packed full of different types of organelles that provide various cellular functions. Different types of cells have differing proportions of organelles, depending upon their function.

The endoplasmic reticulum is a three-dimensional network of tubules that form a mesh within the cell. The endoplasmic reticulum may be either rough or smooth. Rough endoplasmic reticulum is studded with organelles called ribosomes. Ribosomes contain RNA and are the site of protein synthesis. The required amino acids are transported to the ribosomes along the endoplasmic reticulum for assembly into proteins. The proteins are then transported to the Golgi apparatus. The Golgi apparatus is a membrane-bound organelle that packages and transports macromolecules, such as proteins and lipids, throughout the cell. Proteins are packaged in membrane sacs and are then transported to whichever part of the cell requires the protein.

All of the processes that take place within a cell require sources of power. The activities in a cell are powered by organelles called mitochondria. Mitochondria play an essential role in the release of chemical energy from food. A smooth membrane that encloses numerous shelf-like extensions surrounds mitochondria. The number of mitochondria in a cell gives an indication of how active the cell is. Highly active cells have many mitochondria that supply lots of energy, while low-activity cells have few mitochondria, because they need less energy.

Cells also contain a large number of other structures such as microtubules, microfilaments, intermediate filaments, and a cytoskeleton. These structures are used by the cell to aid in reproduction and to give an individual cell its shape. Cells also have storage sacs that hold foods, proteins, amino acids, lipids, wastes, and other cellular products that are stored for either use or eventual disposal. Cells also contain sacs called lysosomes. Lysosomes are important as they contain about 50 different enzymes that can break down organic molecules within the cell. These enzymes are stored in an inactive state surrounded by a membrane. When needed, they are used to break down food molecules or perform other functions. Some cells are genetically programmed to die at a certain point in their life cycle. In these cells, the lysosomes rupture and the enzymes breakdown and digest the cell.

PROKARYOTES AND EUKARYOTES

Scientists now believe that cells of higher plants and animals are descended from much more primitive cells. These simple cells

Measuring in Microns

A micron is a millionth of a meter, so 100 microns is a tenth of a millimeter. The symbol for micron is μm. The term micron is still commonly used in the United States, but it is not a unit recognized by the International System of Units (SI). In the SI system, the equivalent unit is the micrometer.

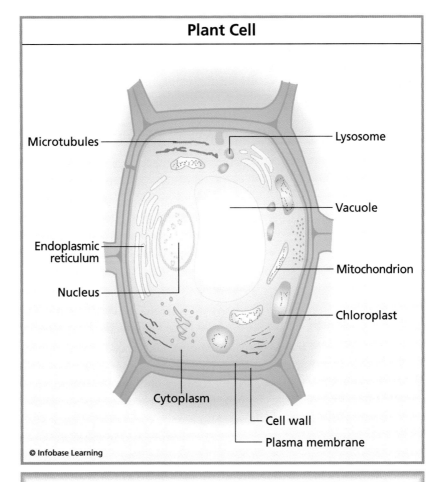

Figure 1.2 A plant cell is eukaryotic, but it differs from the cells of other eukaryotic organisms. Its distinct features include a vacuole; a cell wall composed of cellulose, hemicellulose, pectin, and sometimes lignin; and the chloroplasts, which contain chlorophyll and the biochemical systems for photosynthesis.

with their primitive characteristics still exist today and are called **prokaryotes**. Prokaryotes are small cells, usually from around 1 to 15 microns in size. They have a plasma membrane and a cell wall but they lack a nucleus. They usually have a small, circular piece of DNA that stores genetic material and the coding for proteins.

(continues on page 20)

Endosymbiotic Theory

The endosymbiotic theory concerns the origins of mitochondria and chloroplasts. According to this theory, these organelles originated as separate prokaryotic organisms that were taken inside the cell as endosymbionts. Mitochondria developed from proteobacteria and chloroplasts from cyanobacteria. This theory was first proposed in the 1880s by botanist Andreas Schimper who observed that chloroplasts reproduced in a manner similar to bacteria. The theory was largely ignored until the 1960s. Biologist Lynn Margulis expanded on the theory and included mitochondria as an endosymbiont. Through her work, the theory became widely accepted by mainstream scientists in the 1980s.

Much evidence exists to support this theory. Both mitochondria and chloroplasts contain DNA that is different from that of the cell nucleus and that is circular and similar in size to that of bacteria. Two or more membranes surround mitochondria and chloroplasts, and the innermost of these membranes shows differences in composition from the other cell membranes. The composition resembles the composition of a prokaryotic cell membrane. New mitochondria and plastids are formed only through a process that is similar to reproduction in bacteria.

Even though mitochondria and chloroplasts were once free-living organisms, they are no longer able to exist outside of a cell. Based on these and other observations, scientists are fairly certain that mitochondria and chloroplasts are indeed endosymbionts.

Figure 1.3 The endosymbiotic theory claims that mitochondria and chloroplasts in eukaryotic cells originated from smaller prokaryotic cells living inside larger cells.

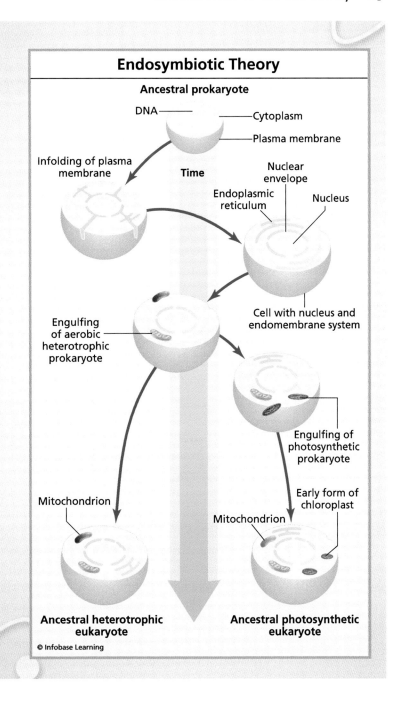

(continued from page 17)

Prokaryotes also lack membrane-bound organelles, so all processes take place only within the cytoplasm. Because of this, prokaryotes cannot carry out the same kinds of specialized functions that more complex cells perform.

Eukaryotes are larger than bacteria. They are usually from 2 to 1,000 microns in size. Eukaryotes are the more familiar type of cells. They have an outer plasma membrane, a nucleus, cytoplasm, and membrane-bound organelles. Because of the organization of membrane-bound organelles, eukaryotes can carry out much more complex and specialized functions.

Animal and plant cells are very similar, and the description of the different parts of cells applies to both of them. Both cells contain an outer plasma membrane, a nucleus, cytoplasm, and organelles. However, plant cells have a few additional types of structures not found in animal cells.

Plant cells have a rigid cell wall in addition to a plasma membrane. The cell wall is outside of the plasma membrane and functions to keep the cell in a rigid shape. The cell wall is composed of cellulose fibers embedded in a matrix of interlinked polysaccharides. The cell wall depends on osmotic pressure to maintain its rigid shape. Osmotic pressure is the pressure that water exerts when the cell wall is filled with water. You have probably noticed that when a plant needs water, its leaves start to droop. When the plant is watered, it "perks" back up and the leaves stop drooping. This occurs because the lack of water in the plant lowers the osmotic pressure exerted on the cell wall, which causes the wall to weaken. The weakening of the cell wall causes the plant to droop.

Plant cells also have another structure called a chloroplast. Chloroplasts are found in algae such as seaweed, some protists, and in plants. The chloroplast is the site where photosynthesis takes place. Chloroplasts use carbon dioxide from the air and water to make glucose using energy from the sun.

Chloroplasts are similar to mitochondria in their structure. Inside the chloroplast are membranes in which the photosensitive pigments chlorophyll and carotenoids are embedded. Chlorophyll is green, which is what gives plants their color. Carotenoids are red, yellow, or orange pigments. Because plant cells contain less carotenoids, they do not affect the color of living leaves. When leaves begin to die during the season of fall, the chlorophyll breaks down and the carotenoids give leaves their fall colors.

STUDYING CELLS

Scientists continue to study cells to unlock more secrets about how they originally formed, how they function, and how they reproduce. Because cells are so small, scientists must use specialized tools to study them. Scientists use different types of **microscopes** as well as microprobes to identify what is happening within the cell.

Tools for Studying Cells

Cells are the structural and functional unit of living organisms. A cell is the smallest unit of an organism that is classified as living. Some organisms, such as bacteria, consist of a single cell. These organisms are called unicellular. Other organisms, such as humans, are multicellular. Humans have an estimated 100 trillion, or 10^{14}, cells. Almost all cells are invisible to the human eye without a microscope, except for some egg cells. For example, a chicken egg is actually a single cell. The largest known cell is an ostrich egg. Eggs are an exception to the rule regarding the normal size of cells. Because all other cells are so small, with most of them ranging from 1 to 100 micrometers, scientists need to use special instruments to observe them.

Most cells are so small, that they are not even visible with a magnifying glass. Therefore, scientists use a magnifying instrument called a microscope. Simple microscopes are low powered and only have a single lens. Compound microscopes use a combination of lenses to magnify tiny objects so that they are visible. Microscopes allow scientists to observe cells and even observe some of their organelles. However, microscopes do not enable scientists to see how cells function and to observe the processes that take place within them. For this knowledge, scientists depend on other instruments.

Tools for Studying Cells 23

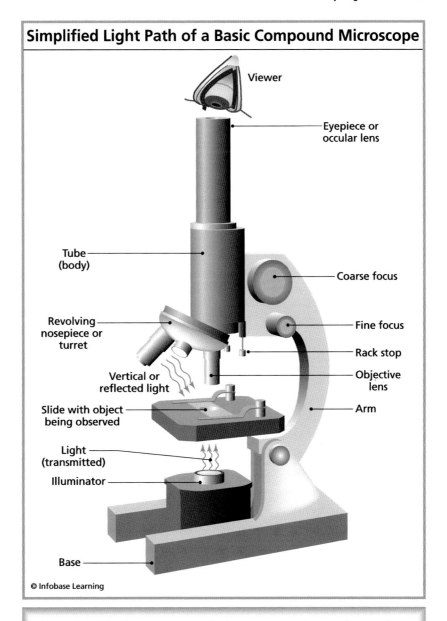

Simplified Light Path of a Basic Compound Microscope

© Infobase Learning

Figure 2.1 The basic compound microscope was invented around 1600 and is still in use today. In this kind of microscope, light shines from below the stage and through a sample. This light is then transmitted through multiple lenses, thereby enlarging an image of the sample for the human eye to see through the eyepiece.

HISTORY AND DEVELOPMENT OF THE MICROSCOPE

The Romans invented glass during the first century A.D. They first experimented with different shapes of clear glass. At some point, someone found that a circular piece of glass that was thick in the middle and thin on the edges could make objects look bigger. Then, someone found that these pieces of glass could focus the sun's rays on parchment paper and set it on fire. The piece of glass was called a lens because of its shape. These pieces of glass were also called either magnifying glasses or "burning glasses." Magnifying glasses remained a novelty until the end of the thirteenth century when wearable eyeglasses were invented in Italy.

The early simple microscopes were merely a tube with a plate for the object at one end and a single lens at the other. They had a power of about 6x to 10x. (This means that they made objects appear 6 to 10 times larger than their actual size, which is about the same power as a common magnifying glass has today.) In fact, these simple microscopes with their single lens were little more than magnifying glasses. They were commonly used to observe small objects. Among the very common and interesting objects that were studied were fleas and other tiny insects. Because of this, early microscopes were often called "flea glasses."

The compound microscope was invented around 1600 in the Netherlands. It is not known who really invented the compound microscope. Three different eyeglass makers have been given credit for the invention: Hans Lippershey, Hans Janssen, and Zaccharias Janssen. Hans Lippershey and Zaccharias Janssen are also credited with inventing the telescope. However, most accounts give credit to Zaccharias Janssen and his son Hans Janssen for inventing the microscope.

Around 1597, Zaccharias Janssen and his son Hans were experimenting with lenses. They observed that placing two lenses in line within a tube magnified close objects. However, their lenses were rather large and the magnification obtained was only about 10x. Italian scientist Galileo Galilei also designed a compound microscope, but it was only useful with reflected light. Robert Hooke (1635–1703) built the first useable British compound microscope around

1655. Hooke used his microscope to examine the structure of cork. (Cork is a plant tissue that comes from the cork oak tree.) Hooke discovered that cork was made up of tiny chambers that he called cells. Hooke's observations were limited by the poor quality of lenses available at the time.

Antonie van Leeuwenhoek (1632–1723), another Dutchman, made lenses as a hobby. Around 1670, he developed a new method for grinding very small glass lenses. The lenses were about 0.03 inches (about 1 millimeter) in diameter and could magnify objects several hundred times. He mounted a lens in a brass plate to make a simple microscope. He then used light to look at objects in a drop of water on the end of a metal pin. A series of screws was used to move the pin from side to side and back and forth to focus the specimen.

Van Leeuwenhoek's microscope design was probably influenced by the designs in Robert Hooke's book *Micrographia.* Van Leeuwenhoek discovered bacteria, sperm, blood cells, and many different protozoa. His work serves as the foundations of plant anatomy. Van Leeuwenhoek reported his discoveries to the Royal Society of London in a series of letters. He made hundreds of microscopes over the years. Today, nine of his original microscopes still exist.

Hooke made a copy of van Leeuwenhoek's microscope because it was much more powerful than his own compound microscope design. Hooke used his new and more powerful simple microscope to confirm van Leeuwenhoek's work. Hooke also confirmed one of his own most important discoveries regarding the cell.

MODERN OPTICAL MICROSCOPES

Other than minor mechanical and optical improvements, no major improvements were made to the compound microscope until the nineteenth century. In 1847, Carl Zeiss began building simple microscopes in Jena, Germany. In 1857, he started building a compound microscope called the *Stand I*. His business grew as he sold more and more microscopes. In 1866, optical glass expert Otto Schott joined the company. Schott was a glass chemist and developed higher quality optical glass for Zeiss. In 1872, Zeiss hired Ernst Abbe. Abbe's

26 CELL THEORY

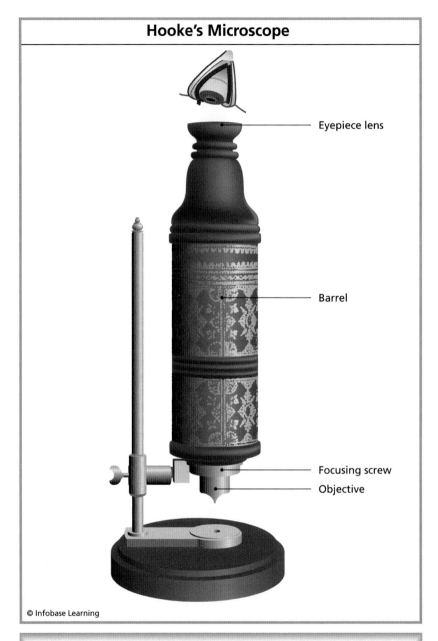

Figure 2.2 The microscope Robert Hooke designed circa 1665 was 6 inches (15.2 cm) long. Although its craftsmanship was said to be excellent, its fragile focusing mechanism would wear out fairly quickly.

work on optical design led to the discovery of many basic facts about optics and lens design. Zeiss soon became known for producing the best optics and microscopes in the world. Compound microscopes were soon being made in many different countries, including in Germany, Great Britain, and the United States. Hundreds of different designs of microscope were developed, most of them in the United States and Great Britain.

Zeiss's company made three great contributions to optics—apochromatic lenses, plan apochromatic lenses, and immersion lenses. Apochromatic lenses are designed to bring three wavelengths (typically red, green, and blue) into focus in the same plane. This is achieved by adding special coatings to the lens. A plan apochromatic lens is actually a system of lenses constructed as two groups of six lenses. An immersion lens is one designed for immersion in oil that has the same light properties as glass. This type of lens allows a much higher magnification.

All modern optical microscopes have the same basic components—eyepiece, objective lens, stage, illumination source, and the body. The eyepiece is the part of the microscope that you look into with your eye. The eyepiece, or ocular, is made up of two or more lenses within a cylinder. The eyepiece brings the image into focus. Eyepieces are usually interchangeable and they may be swapped out to change magnification. Eyepieces are usually 2x, 5x, or 10x.

The objective lens is set at the lower end of the microscope, close to the object being viewed. It is made up of one or more lenses that typically have a magnification of 4x, 5x, 10x, 20x, 40x, 80x and 100x. The 100x lens is an oil immersion lens. To find the total power of the microscope, you multiply the power of the eyepiece by the power of the objective.

The stage is the platform that supports the specimen that is being observed. The specimen is usually placed on a slide, a rectangular piece of glass or plastic. The specimen is placed over a hole in the stage that allows light to pass through. An illumination source lights up the specimen. The illumination source could be sunlight reflected by a mirror through the microscope or it could come from an electric light. Most microscopes have some kind of diaphragm that controls the amount of light that reaches the specimen.

The body of the microscope holds the pieces of the microscope rigidly. Adjusting knobs allow the microscope to focus by moving the objective closer to or farther away from the specimen. There are usually two adjusting knobs, one for coarse focusing and the other for fine focusing. Some microscopes have a place for attaching a camera to record the object under observation.

Many microscopes have three or four objectives set on a rotating turret. A low-power objective, usually a 4x, is called the scanning objective. It is used to find the general area of interest on the specimen. The next two objectives are usually a 10x low-power objective and a high-power 40x objective. If the microscope has a fourth objective, it is usually a 100x oil immersion objective. Modern optical microscopes are usually limited to a high magnification of around 1,000x. Magnifications above that usually suffer from optical problems in the lenses.

ELECTRON MICROSCOPES

Electron microscopes are instruments that use a beam of highly energetic electrons to examine specimens. Electron microscopes have a much greater resolution power than light microscopes. This means that electron microscopes are much more powerful than light microscopes. While most light microscopes are limited to about 1,000x magnification, electron microscopes have up to 2 million times magnification.

Electron microscopes function in a similar way to optical microscopes except that a focused beam of electrons is used instead of light. Basically, a stream of electrons is formed by an electron source and is directed toward a specimen by using a positive electrical field. The electron stream is narrowed and focused by metal apertures that restrict the size of the beam and magnetic fields that act like lenses to focus the beam. Once the beam is focused on the sample, interactions occur between the sample and the electron beam. This disturbs the electron beam and that disturbance is used to create an image. The exact nature of how disturbances in the electron beam are converted into an image depends on the type of electron microscope.

German scientists Max Knoll and Ernst Ruska invented the transmission electron microscope (TEM) in 1931. Then, Ruska alone

Tools for Studying Cells 29

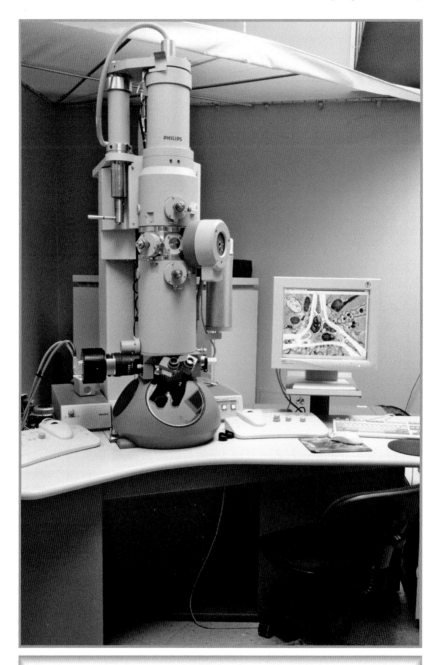

Figure 2.3 A transmission electron microscope (*seen here*) operates much like a light microscope, but it uses electrons instead of light. As a result, electron microscopes give a much higher magnification—about 1,000 times better—than light microscopes.

improved upon the design in 1933. The TEM uses a high voltage electron beam emitted by a tungsten filament and focused by magnetic lenses. The electron beam is transmitted through an ultrathin specimen that is partially transparent to electrons and, thus, partially scatters the beam. When the beam emerges from the specimen, it carries information about the inner structure of the specimen. The variation among the unaffected electrons and scattered electrons in this information, or the "image," is then magnified by a series of electromagnetic lenses until it is recorded by hitting a fluorescent screen, photographic plate, or light sensitive sensor such as a digital camera. The image detected by the sensor may be displayed in real time on a monitor or computer and recorded for later viewing.

In 1935, Max Knoll built the first scanning electron microscope (SEM). Unlike the TEM, where electrons of the high voltage beam form the image of the specimen, the SEM produces images by detecting low energy secondary electrons that are emitted from the surface of the specimen due to excitation by the primary electron beam. In the SEM, the electron beam is scanned across the sample in parallel lines. Detectors build up an image by comparing the signals with the beam's position to create a map. Yet, generally, the TEM resolution is about an order of magnitude greater than the SEM resolution, because the SEM image relies on surface processes rather than transmission. Therefore, the SEM is able to image bulk samples and has a much greater depth of view, and so it can produce images that are a good representation of the three-dimensional structure of the sample.

In 1981, Gerd Binnig and Heinrich Rohrer invented the Scanning Tunneling Electron Microscope (STEM) in Switzerland. The STEM uses a focused beam to scan across a specimen that has been prepared in a thin section to facilitate detection of electrons scattered through the specimen. STEM is useful because it combines the high resolution of TEM and the spread-out beam of SEM. As a result, the STEM has the advantages of both types of electron microscopes.

Electron microscopes are used for different purposes than light microscopes. Electron microscopes are used to study topography, morphology, composition, and crystallographic information. Topography refers to how a surface looks. This includes its texture and its material properties. The morphology refers to the size and shape of the particles that make up a specimen. The composition refers to the chemicals that make up a specimen. The crystallographic information tells how the atoms and molecules in a specimen are arranged. All this is much different from that information provided by light microscopes.

Electron microscopes are not without their disadvantages. They are very expensive to buy and maintain. Plus, they require very stable, high-voltage electricity, an ultra-high-pressure vacuum system, and a cooling water system. Furthermore, they require a very stable platform as they are extremely sensitive to vibrations and external magnetic fields. Most high-powered electron microscopes are housed in separate buildings or underground rooms. All of these requirements make it expensive to use and maintain an electron microscope.

Electron microscopes also require that specimens be prepared with elaborate procedures before being observed. First, the specimen is frozen in liquid nitrogen or liquid helium. Then, the specimen is dehydrated by replacing any water with an organic solvent such as ethanol or acetone. Next, the specimen is embedded in epoxy. The epoxy block is then cut into thin sections and polished to a

Nobels for Microscopes

In 1986, half of the Nobel Prize in physics was awarded to Ernst Ruska for his work with electron optics and the electron microscope. The other half of the Nobel Prize was awarded to Gerd Binnig and Heinrich Rohrer for their design of the scanning tunneling microscope. The invention of the light microscope was a major advance in the study of cells, but it was limited by the optics and the transmission of light through those optics. The electron microscope designed by Ruska was a major step forward as it created more resolution than the light microscope.

Gerd Binnig and Heinrich Rohrer made advancements in electron microscopy with the development of the scanning tunneling electron microscope. They overcame technical issues, enabling them to develop the STEM. Their groundbreaking work opened up the field of electron optics.

smooth, mirror-like finish. The specimen is then coated or stained with a heavy metal such as lead, uranium, tungsten, platinum, or gold. Finally, the specimen is ready for observation in a vacuum chamber within the electron microscope. A vacuum chamber is required because molecules in the air would cause the electrons in the electron beam to scatter.

OTHER TOOLS FOR STUDYING CELLS

Cells are difficult to study because they are very small. In addition to microscopes, scientists have a few other tools that aid in the study of cells. One method that is used to study individual organelles is to disrupt the cells and extract the specific organelle of interest. Once the specific organelles are extracted, scientists can do various things such as extract the proteins that make up the organelle and look at the individual organelles with a microscope. With this information, scientists can determine how the organelles are made, what they are made of, and how they function. This technique does yield information, but much of it involves trying to figure out the answer from many different pieces. Computer models are often used to help scientists figure out exactly how the organelles function.

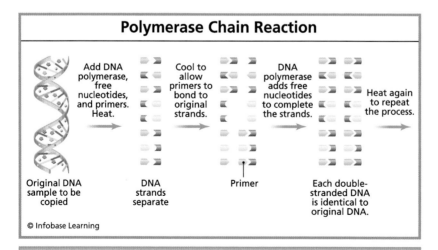

Figure 2.4   The polymerase chain reaction can theoretically increase the amount of a DNA sequence by more than 1 billion times.

DNA and RNA in cells are studied by using techniques such a polymerase chain reaction (PCR). PCR is a technique that is used to amplify the number of copies of a specific region of DNA in order to produce enough DNA to be adequately tested. To use PCR, one must already know the exact sequences or markers that lie on either side of both ends of a given region of interest in DNA. This may be a gene or any sequence. It is *not* necessary to know the DNA sequence in-between the markers. The building-block sequences (nucleotide sequences) of many of the genes and flanking regions of genes of many different organisms are known.

The microscope is still the main tool that scientists use to study cells. The history and development of the microscope was important in helping scientists gain knowledge about cells. Because scientists still use this same basic tool, they rely upon advancements in optics and technology to learn more than their predecessors knew.

Developing the Cell Theory

The development of the cell theory took place long after cells were first discovered. Part of the reason for the delay was that microscope technology failed to make any progress. In fact, some of the early microscopes built by Antonie van Leeuwenhoek had better optical properties than later microscopes. Big advancements in the study of cells did not take off until the 1800s.

THE DISCOVERY OF CELLS

Because most cells are too small to be observed with the naked eye, their existence was unknown to science until the late 1600s. Antonie van Leeuwenhoek is credited with being the first to discover cells. Van Leeuwenhoek was born in Delft, Holland. As a teenager, he worked as an apprentice to a cloth manufacturer. When his apprenticeship ended, he was promoted to head bookkeeper due to his methodical work habits. Because he worked with woven fabrics, he was interested in finding new ways to assess their quality. He learned of the invention of the single-lens microscope and decided to build one of his own for looking at fabric. Over the years, his interest in both microscopes and nature led him to start studying nature. He never received any formal training in the sciences and worked solely on his own time after working all day. One of his big discoveries came when he studied water. He discovered single-celled organisms, which he called "animalcules," the organisms that today we call bacteria and protists. He also turned his microscope onto larger organisms such as lice and bees to study the details of their mouthparts and stingers.

Figure 3.1 Antonie van Leeuwenhoek, the Dutch pioneer of microscopic research, provided the first accurate description of blood corpuscles, spermatozoa, and microbes.

Van Leeuwenhoek's work caught the attention of the Dutch biologist Regnier de Graal. De Graal contacted the Royal Society of London in 1673 and told them of van Leeuwenhoek's work. Van Leeuwenhoek was then contacted and eventually published his observations in the *Philosophical Transactions of the Royal Society of London*.

Van Leeuwenhoek made many contributions to science and biology. However, his greatest contribution to the cell theory, other than his discovery of single-celled organisms, was his work on improving the microscope. He discovered a way of taking several lenses and placing them in a holder, thus creating the compound microscope. The combination of lenses in the holder did not significantly increase magnification, but it did produce a much clearer image. He tinkered with the design of microscopes and eventually produced one that provided about 300x magnification. After his death, he willed all of his microscopes—numbering more than 400—to the Royal Society of London.

Englishman Robert Hooke expanded on van Leeuwenhoek's observations with the newly developed compound microscope. Hooke used his compound microscope to observe many different objects. One of them was a piece of cork. While studying the cork, he noticed that the tiny, boxlike compartments in the cork resembled the cells of a monastery. The term *cell* was born. However, he viewed cells simply as a container, not as the basic unit of life.

Hooke also reported seeing similar cell structures in wood and in other plants. In 1678, the Royal Society of London asked Hooke to review the findings of van Leeuwenhoek. Hooke looked at the tiny "animalcules" and confirmed van Leeuwenhoek's findings. Hooke's work gave credibility and acceptance to van Leeuwenhoek's discoveries.

Even though Hooke is considered the father of microscopy, little is known about him. There are no known portraits of him. He had a falling out with one of his colleagues, Sir Isaac Newton, who worked hard to suppress all of Hooke's work. Hooke's interests covered almost all of the fields of science and technology—physics, astronomy, chemistry, biology, geology, architecture, and naval technology. Hooke collaborated or corresponded with most of the great scientists of his time, such as Christian Huygens, Christopher Wren, Robert Boyle, and Isaac Newton.

Figure 3.2 British scientist Robert Hooke made numerous discoveries in fields as diverse as astronomy and microbiology, yet there are no portraits of him. Thus, this bust—from the Hooke Museum on the Isle of Wight in the United Kingdom—was based only on written descriptions of him. While there is some controversy over why this is the case, some historians suggest that fellow British scientist Isaac Newton, who died 24 years later, attempted to eradicate Hooke's likeness from history due to their long rivalry over studies in light and gravitation, amongst other things.

Hooke was responsible for many inventions and discoveries. He invented the universal joint, the iris diaphragm, and an early prototype of the respirator, as well as the anchor escapement and the balance spring, which made more accurate clocks possible. He also invented or improved meteorological instruments, such as the barometer, anemometer, and hygrometer. Hooke helped worked out the correct theory of combustion. He also devised an equation describing elasticity that is still used today as Hooke's law. He made many contributions to many different fields, but his work in biology really stands out.

Hooke's first major published work was *Micrographia*, one of the most significant works ever published in biology. His book set the foundation for using microscopy as a tool in biology. *Micrographia* was published on November 23, 1665, and became a bestseller. Some critics were amazed by the book and its detailed drawings, while others ridiculed Hooke for wasting his time on such a trivial pursuit. The book's illustrated pages represent Hooke's 30 years of achievements with his microscope. Hooke's observations were recorded and accompanied by detailed drawings. His work inspired many scientists to study and expand on his work.

PROGRESS BETWEEN 1700 AND 1838

With the invention of the microscope and the early work by van Leeuwenhoek and Hooke, scientists began looking at the world through microscopes. Most scientists at the time studied plants. They began exploring the microscopic structure of plants and tried to figure out how they work. Advances in both the microscope and staining techniques to identify cell parts played a huge role in a number of discoveries. The work of many scientists during this time lead to the development of the idea that cells were actually the basic structure of life—both plant and animal life.

The concept of cells began to spread throughout biology. As mentioned, most of this work was done on plants because they were easy to section and observe. Caspar Friedrich Wolff (1733–1794), a German physiologist who was one of the founders of embryology, theorized that tissue was a homogeneous matrix filled with bubbles, as is rising dough. He postulated that organs are formed in differentiated layers from undifferentiated cells. This went against the conventional thought at the time because it was widely believed that organisms were created as fully-formed from miniature versions of themselves that existed since the beginning of creation.

The work of Wolff was built upon by French botanist Charles-François Brisseau de Mirbel (1776–1854). Mirbel founded the fields of cytology (the study of cells), plant histology, and plant physiology. He proposed that all plant tissue is modified from parenchyma. In plants, parenchyma is the primary tissue of higher plants. It is composed of thin-walled cells and forms the greater part of leaves, roots, fruit pulp, and the pith of stems. In 1809, Mirbel observed that each plant cell is contained within a continuous membrane. This idea remains a central contribution to cytology.

The idea that cells were separable into individual units was proposed by Ludolph Christian Treviranus (1779–1864) and Johann Jacob Paul Moldenhawer (1766–1827). Treviranus discovered intercellular spaces and that the epidermis of plants was made up of individual cells. Moldenhawer discovered a way to separate plant cells from the middle lamella layer that separates them. He found that plants have vascular and parenchymatous tissues, which each have different functions. Then he identified vascular bundles and the cells that make up the cabium, and he observed that the stomata on the underside of leaves are composed of pairs of cells instead of a single cell with a hole. Although Moldenhawer is not credited with developing the cell theory, his work provided key documentation for the validity of the theory in plants.

In 1832, Barthelemy Dumortier (1797–1878) of France first described "binary fission," or cell division, in plants. He observed the formation of a mid-line partition between the original cell and the new cell. His observations led him to reject the idea that new cells arise from within old ones or that they form spontaneously from noncellular material.

Franz Julius Ferdinand Meyen (1804–1840) was a German physician and a botanist. He travelled extensively and collected and described many plants. In 1834, he wrote the textbook *Phytotomie*, in which he described a cell theory where new plant cells arise through cell division rather than free cell formation. His theory categorized the plant organs known as Merenchym, Parenchym, Prosenchym, and Pleurenchym according to the different form of their cells and suggested that the growth of the plants is due exclusively to cell division.

René Joachim Henri Dutrochet (1776–1847) was a French physician, botanist, and physiologist. He was the first to observe the diffusion of a solvent through a semi-permeable membrane, a process he called osmosis. He also connected the green pigment in plants and the use of carbon dioxide to plant metabolism. In 1824, Dutrochet performed many microscopic examinations of tissues of both plants and animals and found that both were made up of cells. Dutrochet was among the first to recognize the importance of an organism's individual cells and stated, "It is clear that it constitutes the basic unit of the organized state; indeed, everything is ultimately derived from the cell."

In the early 1830s, Scottish naturalist Robert Brown (1773–1858) reported the discovery of the cell nucleus. While studying orchids, he noticed that the epidermis cells had opaque spots, which he called the areola or nucleus. Other scientists had likely seen these opaque spots but it was Brown who actually recognized them as a part of the cell and gave them a name. He then found this same spot in almost all the plant cells that he observed. Brown recognized that the nucleus was important, but he did not suggest a potential function for it.

SCHLEIDEN AND SCHWANN

German botanist and physician Matthias Schleiden (1804–1881) expanded on Brown's discovery of the nucleus. Observations that Schleiden made in 1832 led him to conclude that the cell nucleus must somehow be connected with cell division. However, he mistakenly believed that new cells erupted from the nuclear surface like blisters. Even so, he made other accurate observations about plant cells and cell activity and his conclusions marked the beginning of plant cytology. By 1837, his work led him to propose the theory that every structural element of plants is composed of cells or cell products.

In 1838, Schleiden was discussing plant and animal cells and his observations of the nucleus while dining with colleague Theodor

Developing the Cell Theory **41**

Figure 3.3 German physiologist Theodor Schwann defined the cell as the structural unit of all living organisms in 1839.

Schwann (1810–1882). Schwann was a German physiologist who worked with animal cells. He had seen similar structures in the cells of the notochord. The two quickly went to work and confirmed these observations, as well as making new ones. Schwann and Schleiden recognized that some organisms are single-celled while others are multicellular. They also recognized that membranes, nuclei, and cell bodies were common cell features and described them by comparing various animal and plant tissues. Schwann published *Microscopic Investigations on the Accordance in the Structure and Growth of Plants and Animals* in 1839. In this work, he made the following observations:

- Cells are organisms and all organisms consist of one or more cells.
- The cell is the basic unit of structure for all organisms, and plants and animals consist of combinations of these organisms, which are arranged in accordance with definite rules.

In other words, the cell is the basic unit of life. This statement, along with Schleiden's statement, established the cell theory. Both scientists also concluded that cells form by free-cell formation, similar to the formation of crystals. In other words, they form by spontaneous generation. However, this part of the cell theory was incorrect and would later be replaced.

RUDOLF VIRCHOW

Rudolf Ludwig Karl Virchow (1821–1902) was a German physician, anthropologist, public health activist, pathologist, biologist, and politician. He is considered the "father of pathology." Virchow worked in many fields and made many important discoveries. One that stands out is his addition to the cell theory. Even though Francesco Reddi (1626–1697) had performed the famous experiments with meat and maggots to prove that spontaneous generation did not exist for animals, the concept of spontaneous generation was still widely accepted in the study of cells. Even Schleiden and Schwann did not dispute spontaneous generation in their cell theory. Reddi's work gave rise to the maxim *Omne vivum ex ovo* (every living thing comes from a living thing). In 1858, Virchow extended this by stating that the only source for a living cell was another living cell. His statement replaced the spontaneous generation statement found in Schleiden and Schwann's cell theory. Virchow's addition gave rise to the classical cell theory:

- All organisms are made up of one or more cells.
- Cells are the basic unit of life and structure.
- All cells come from pre-existing cells.

Additional confirmation of Virchow's addition to the cell theory came from Louis Pasteur (1822–1895). Pasteur set out to disprove spontaneous generation with a now-classic experiment that both firmly established the cell theory beyond doubt and solidified the

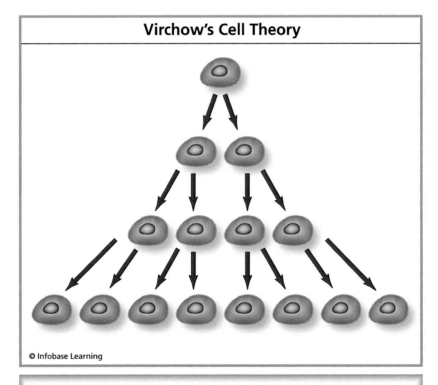

Figure 3.4 Rudolf Virchow's theory that every cell originates from a similar, previously existing cell applies equally true in the cases of tissues of the eye, a maggot that suddenly appears on rotting meat, and a tumor cell.

Figure 3.5 Despite Rudolf Virchow's stellar reputation in the fields of pathology and public health, he strongly rejected the idea that bacteria caused disease.

basic steps of the modern scientific method. In his experiment, he exposed boiled broth to air in vessels that contained a filter. The filter prevented all dust particles from passing through and reaching the broth. He also set up vessels with long winding tubes that kept dust particles from reaching the broth. Pasteur found no organisms growing in the broths, thereby showing that organisms that had been previously found must have come from outside. This meant that the organisms grew from spores from outside dust and did not spontaneously generate in the broth. This was one of the last and most important experiments that disproved the theory of spontaneous generation.

The establishment and acceptance of the cell theory opened the door for more advancement in the fields of cytology, histology,

Spontaneous Generation

The idea of spontaneous generation is usually credited to Greek philosopher Aristotle (384 B.C.–322 B.C.) and his theory of **abiogenesis**. According to Aristotle, life could arise in three different ways—from sexual reproduction, asexual reproduction, or nonliving matter. The idea of spontaneous generation is supported by everyday (but incomplete) observations, such as insects and worms appearing from rotting meat, frogs from mud, and mice from rotting wheat. However, not everyone agreed with this concept. From the time of Aristotle until Louis Pasteur's experiment, the debate over spontaneous generation raged among scientists. From the 1600s to the 1800s, many scientists performed experiments to either prove or disprove spontaneous generation. It was not until Pasteur came along that a convincing experiment was devised.

While Pasteur convinced the scientific community that spontaneous generation does not occur, the debate still continues. Even in modern times, scientists argue about how life first arose from nonliving molecules. This debate will be described in Chapter 7.

pathology, botany, and zoology. The cell theory quickly became accepted because it explained many observations. Many scientists tied their work into the cell theory, and this served to make the cell theory one of the unifying concepts of biology. Work on the cell theory has added more tenets to the theory, but the three classical tenets proposed by Schleiden, Schwann, and Virchow have remained unchanged since they were first proposed.

The Neuron Theory

Ancient Greeks considered the brain to be an empty shell in which "the phlegms"—thought to cause sluggishness, apathy, and evenness of temper—originated. They also believed that the heart was the source of all emotions. Roman physician Galen (199–130 B.C.) was the first to produce an accurate description of the brain's anatomy. He described the brain as having three liquid-filled ventricles. Galen thought that these ventricles were where the "vital breath" of the human body was stored. He further suggested that the solid portions of the brain were where the soul and cognitive functions originated. These ideas of the Greeks and Romans continued to have an influence on thought through the Dark Ages when anatomical studies were ignored and even shunned. By the fifteenth century, it was generally believed that the brain and cerebrospinal fluid acted as a hydraulic fluid to move the arms and legs. The pump for this system was the fluid-filled ventricles in the brain. This idea was defended by both Leonardo da Vinci in the fifteenth century and René Descartes in the sixteenth century.

Before the invention of the microscope, the delicate structures of the brain and nervous system remained a mystery. However, even after the invention of the microscope, examinations of the brain and nervous tissues only presented researchers with more mysteries. With the microscope, they could see a network of very thin, complex structures. It was widely assumed that the nervous system was an interconnected mesh, much like wires that carried nerve signals between the brain and the body. When the cell theory was proposed, these researchers were quick to point out that the nervous system was an exception to the cell theory.

STRUCTURE AND FUNCTION OF NERVE CELLS

Today, scientists understand the structure and function of nerve cells or **neurons**. A nerve cell is made up of three parts: cell body, dendrites, and an axon. The cell body is the main part of the nerve cell. It holds the nucleus and most of the organelles that carry out cell functions. The structure of nerve cells differs from other cells in that they have dendrites and axons. The dendrites are the parts of a nerve cell that are receptive to some stimuli. They collect the stimuli signal and transmit it to the body of the nerve cell. The axon is the long part of the nerve cell that carries the signal over long distances through the body. Each nerve cell has only a single axon. The axon transmits the nerve signal down its length to other nerve cells or to muscles or glands. A single nerve cell only transmits a signal in one direction. Nerve cells are not in direct contact with each other; instead, there is a small gap between the end of the axon and the next dendrite. The nerve impulse passes over this gap chemically with the help of a substance called a neurotransmitter.

Neurotransmitters are chemical molecules that are released by the far end of an axon and travel across the microscopic gap to a dendrite of the next nerve cell. This stimulates the next nerve cell and the signal is then passed along that nerve. The signal is passed along the nerve by a complex process that involves sodium and potassium. As the ions of sodium and potassium, which are charged atoms, move across the cell membrane that surrounds the nerve cell, an electric charge is passed along the surface. This process occurs very rapidly and can move the signal along the axon very quickly. Because nerve cells transmit an electrical signal, they must be insulated just like a wire. When many nerve cells are close together, they form bundles. Each axon is surrounded by a myelin sheath. The

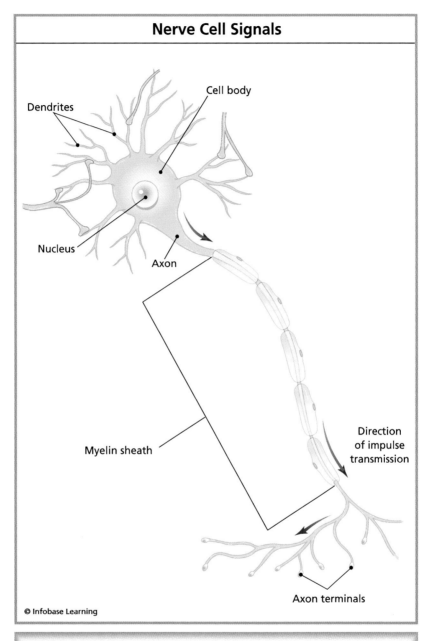

Figure 4.1 Nerve cells have a long axon that is very delicate. Because nerve cells are so fragile, early scientists had a difficult time studying them.

myelin sheath acts as insulation to isolate the nerve cell from other nerve cells. Taken together, all the nerve cells in a body make up the nervous system.

The nervous system varies in complexity depending on the animal. Single-celled organisms do not have a nervous system. Multicellular organisms have specialized nerve cells and need to be able to transmit signals from one part of the organism to another. Simple organisms such as worms have a simple nervous system while humans and most other mammals have very complex nervous systems. In mammals, including humans, the brain serves as the control center for all body functions. The nervous system connects the brain to all parts of the body. In mammals, the nervous system is divided into two parts, the central nervous system and the peripheral nervous system. The central nervous system is made up of the brain and spinal cord, and the peripheral nervous system is made up of all the nerve cells that connect the central nervous system to the muscles, organs, and glands. The human nervous system is composed of an estimated 10 billion nerve cells.

Nerve cells are further classified according to their function. Sensory neurons have receptors that receive a stimulus from the external environment or from within the body. The skin has many such receptors so that a person can feel even a light touch. Motor neurons control the organs that stimulate a physical response. This includes muscles, glands in the digestive system, and glands in the excretory system. Intermediate neurons, or interneurons, are typically found in the central nervous system where they carry signals to and from the brain. Intermediate neurons may also connect the sensory and motor neurons. The neurosecretory neurons are specialized neurons that synthesize and secrete hormones.

The nervous system also contains another type of cell called the glia or neuroglia. (The name is derived from the Greek word for "glue.") The neuroglia is the connective tissue that provides structural support for the nervous system. It is estimated that for every neuron, there are at least 10 neuroglia cells. A neuroglia is much smaller than the neuron, yet they occupy about 50% of the total volume of nerve tissue. Neurons cannot exist or develop without neuroglia. The neuroglia do not conduct any nerve impulses, but they are responsible for forming the myelin sheath around the axons of nerve cells.

Neurons do not divide and only rarely regenerate in higher animals. As a result, damage to nerve cells is often permanent. The neuroglia cells do divide and replace themselves and can repair the connective tissues if they are damaged.

Today, scientists have a good understanding of the structure and function of the nervous system. Much of this understanding was reached at about the same time as the development of the cell theory. The cell theory provided a good explanation for cells in plants, animals, and humans, but it did not seem to explain the nervous system. The early researchers did not believe that the cell theory applied to nerve cells. Only after much work by many different scientists did an understanding of nerve cells arise.

RESEARCHING AN EXCEPTION TO THE CELL THEORY

Nerve tissue is very soft and fragile. Before the microscope was invented, it was impossible to see or trace the path of nerve cells. After the invention of the microscope, it became somewhat easier to see the nerve cells, but the long axons were confusing and very difficult to follow. Early researchers carefully dissected animals using microscopes and needles to carefully follow the nerve pathways. Because of the complexity of the nervous system, and the fact that delicate nerve tissue deteriorates rapidly, it was impossible to construct a three-dimensional model of the nervous system or understand how it functioned. The gap between nerve cells was very tiny and often obscured by the dendrites and the ends of axons. This led early researchers to believe that the nervous system was actually a connected network of tiny fibers that acted like wires to carry signals. Because the nervous system was thought to be interconnected, it did not make sense that it was made up of individual cells because nerve cells did not resemble any other cells in plants or animals. The function of the nervous system was described by the network, or reticular, theory because it was all thought to be interconnected. Because of this, researchers believed that the nervous system did not adhere to the cell theory and so was an exception to it.

52 CELL THEORY

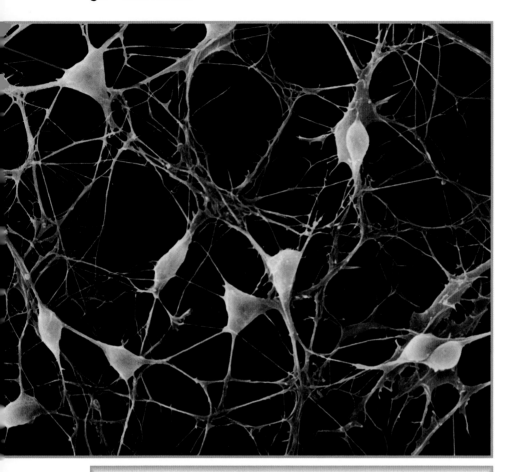

Figure 4.2 Nerve cells appear to have many connections, giving the impression of an interconnected network. Each one has a large cell body with several long processes extending from it, usually one axon (red) and several thicker dendrites. The axon carries nerve impulses away and makes contact with other nerve cells and with muscles or glands. The highly branching dendrites receive information from other nerve cells. This complex network forms the nervous system, which relays information rapidly through the body.

EARLY STUDY OF NEURONS

In 1865, a German-language book describing the nervous system was published, two years after the death of its author Karl Deiters (1834–

1863). The book contained very detailed drawings and descriptions of the nervous system. Deiters performed dissections with a microscope. He made drawings of nerve cells that showed them as having a distinct body, dendrites, and a single, long axon with no branching. He worked with motor neurons from the adult spinal cord. His work was used by other researchers, but it was still widely believed that the nervous system was actually an interconnected network.

One of the researchers who took an interest in Deiter's work was a Swiss professor of anatomy and physiology named Wilhelm His (1831–1904). Wilhelm His studied the developing central nervous system of human embryos. He identified all the parts that would eventually make up the central nervous system and determined their functions. He determined that the nervous system was actually made up of individual cells. His's work was continued by Spanish scientist Ramón y Cajal (1852–1934).

Cajal performed a technique developed by Camillo Golgi (1843–1926), using a silver to stain nerve cells. His early work at trying to unravel the mysteries of the nervous system proved very frustrating. He had started his studies of the nervous system by using adults. However, he found it too complicated to trace out the pathways of the adult nervous system. He then switched to studying the developing nervous system of embryos. He focused on the cerebellum, a part of the brain that is the sensory center. This approach proved to be much more successful. Amazingly, he worked out the development and structure of the cerebellum within two years.

In 1891, Heinrich Wilhelm Gottfried von Waldeyer-Hartz (1836–1921) published a paper that established the neuron theory. His paper was not based on his own research but was instead a synthesis of work done by His, Cajal, and several other scientists. The paper established the neuron theory with the following tenets:

- Axons all arise from nerve cells. There is no connection with a fibrous network, and no origin from such a network.
- All of these axons end freely, with terminal arbours and no networks or anastomoses.

These were aligned with the cell theory and established that the nervous system was indeed made up of individual cells. The cells of the nervous system also followed the cell theory.

> **The Neuron Doctrine**
>
> - The neuron is an anatomical unit.
> - The law of dynamic polarization: Cajal formulated this concept, by which he meant the essentially unidirectional flow of information within neurons traveled from their receptive surface (commonly the dendrites), through or past the cell body to the axon, and hence to its terminal branches.
> - The neuron is an embryological or developmental unit. Cajal showed that the axon and dendrites of a neuron grow out from the cell body during development and are always unconnected.
> - The neuron is a metabolic unit. The distal part of the nerve degenerates following nerve damage.
> - The neuron is a basic information processing unit.

THE NEURON DOCTRINE

Work has continued on neurons in attempts to further understand how the nervous system functions. The resulting work has expanded on what is called the neuron doctrine. The current neuron doctrine is based on five tenets (listed in the sidebar). In the years since the neuron doctrine was first proposed, advances in technology have given researchers new opportunities for studying nerve cells. As a result, more and more information is being learned about nerve cells. This has also led to the addition of further points to the neuron doctrine. Some current researchers consider the neuron doctrine to be too restrictive and have offered proposals to rewrite it in light of new studies. Regardless of the direction that the neuron doctrine takes, it is likely that it will always support the cell theory.

DNA and the Cell Theory

One of the tenets of the cell theory is that all cells come from pre-existing cells. This tenet implies that cells must be able to reproduce. Scientists had long observed cells reproducing by binary fission. As scientists tried to figure out how cells actually managed to reproduce, a question arose: How did the instructions pass from the parent cell to the daughter cell? There had to be a mechanism somewhere within the cell for passing on information.

The cell nucleus contains very long threads of nucleic acids and proteins that are called chromosomes. In the early 1900s, scientists hypothesized that chromosomes carried hereditary information. As work continued in this area, this hypothesis gradually became accepted by biological scientists. Continued work revealed that material within the nucleus indeed contains hereditary information that directs the growth, development, and activities of the cell.

CHROMOSOME THEORY

The concept of the chromosome theory of heredity is very important to the understanding of evolution and genetics. Much research by famous scientists such as Jean-Baptiste Lamarck (1744–1829) and Charles Darwin (1809–1882) set the stage with the theory of evolution and how organisms pass their traits to their offspring. Darwin suggested a hypothesis to explain how traits were passed on and called it pangenesis. The hypothesis described the way in which heritable units are transferred from generation to generation as follows: Throughout development, a multicellular organism's component cells disperse gemmules throughout the body, which develop into cells like their predecessors. These gemmules are collected in the reproductive organs due to their mutual affinity for other gemmules. When two of these gemmules combine, a new organism is created. Some combinations of the elements of some of the gemmules in the packets develop into a new organism. This hypothesis leaves open the idea of inheriting acquired characteristics because the component cells can change due to the environment. Darwin's hypothesis was never widely accepted among scientists.

August Weismann (1834–1914) proposed that sexual reproduction itself was an important factor in the passing of traits to offspring. He proposed the germ plasm theory to explain how this worked. This theory suggested that genetic material was passed through a chemical or a molecular substance, which Weismann referred to as *germ plasm*. The main importance of Weismann's theory is that it asserts that acquired characteristics are not passed to the next generation, leaving natural selection as the exclusive explanation for biological evolution.

For this theory to succeed, it was important to realize that **germ cells** are different from other somatic cells. This realization occurred in 1887, when Walther Flemming observed that the production of sperm was different from mitosis in normal cells. The chromosomes of sperm went through what looked like normal mitosis, but then the second division resulted in half the number of chromosomes. This was the evidence that Weismann needed to confirm his theory. He suggested that because germ plasm has been transmitted from generation to generation all along, and because two germ plasms are combined in sexual reproduction, each germ plasm must be halved. Weismann never really said that the hereditary material is located on chromosomes. However, he did believe that the nucleus of the germ cells has the ability to transfer hereditary material.

Austrian monk Johann Gregor Mendel (1822–1884) performed breeding experiments with garden peas. His careful note taking and

DNA and the Cell Theory 57

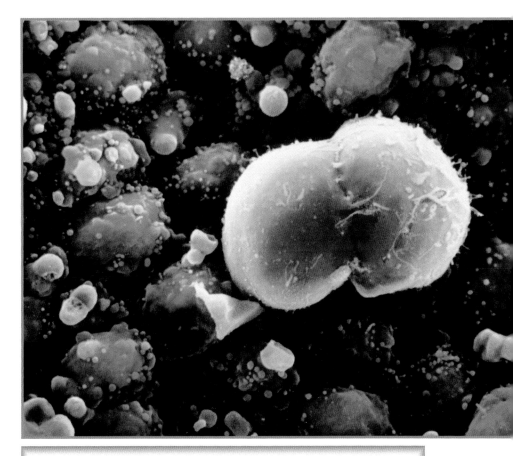

Figure 5.1 This false-color scanning electron micrograph shows a germ cell undergoing mitotic cell division in the ovary of a seven-week-old embryo. At this stage, germ cells migrate into the ovarian cortex where they multiply by mitosis.

work led him to formulate the basic laws of heredity. Mendel published his findings in 1866, but his discoveries were ignored until 1900 when a number of researchers independently rediscovered Mendel's work and grasped its significance. Mendel's work soon became known as Mendel's law. His work conclusively proved that Darwin's theory of pangenesis was wrong.

Mendel used the pea plant to show that hybrids form different kinds of pollen and egg cells. The resulting variation in offspring is due to the segregation of particulate factors when sex cells are

Figure 5.2 Austrian scientist Gregor Mendel raised and studied pea plants. He crossbred plants with different traits and kept track of all the offspring. His work formed the basis of modern genetics.

formed. Mendel found that the segregation of heritable factors follow a constant law. He also noted that there are both dominant and recessive traits. He also noted that there is no blending of traits, rather the hybrid will have parental traits based on a fixed scheme. He noted that traits expressed in offspring followed a statistical pattern. While Mendel worked out the statistics of inheritance, his work did not identify what structure carried the traits.

The first proof that chromosomes carried hereditary material first came from Walter Sutton (1877–1916), who studied grasshopper cells. He observed that during meiosis, chromosome pairs split, making two daughter cells. This division of chromosomes in sex cells could explain Mendel's law.

Thomas Hunt Morgan (1866–1945) was critical of the theories of both Darwin and Mendel. He began working with the fruit fly *Drosophila melanogaster* in 1908. His intended goal was to distinguish Darwinian and Neo-Lamarckian evolutionary theories by using the same type of statistical analysis that Mendel used with pea plants. Morgan used the new information on chromosomes, combining Mendel's law with the knowledge of chromosome division.

Morgan became convinced that chromosomal differences in sperm and eggs correlated with sex determination and concluded that both eggs and sperm played a role in determining sex. He later modified this idea by saying that the heredity of sex can be best understood when one sex is regarded as the pure line, or homozygous, and the other sex is treated as heterozygous. He determined this by crossing a white-eyed male fly with a red-eyed female. The offspring were all red-eyed, but the second generation was 75% red-eyed and 25% white-eyed. Furthermore, the white-eyed flies were all male, and there were twice as many red-eyed females as red-eyed males. This proved that eye color was a typical Mendelian trait and that indeed the sperm did carry sex factors. Establishing the linkage between sex and another non-sex trait, along with further experiments, showed that sex is determined by chromosomes, and the gene for eye color is on the same chromosome. Morgan was able to show that genes are present in alleles that have a specific order on the chromosome. This work proved conclusively that traits were carried on chromosomes.

THE DISCOVERY OF DNA

In April 1953, researchers James Watson (1928–) and Francis Crick (1916–2004) wrote, "This structure has novel features which are of considerable biological interest." The structure they were writing about was deoxyribonucleic acid, or DNA. That humble sentence did not even begin to sum up the importance of their work. Perhaps a better description of their work was provided by Francis Crick, just one month earlier in a pub in England, when, while socializing with several of his colleagues, he said, "We have found the secret of life." Watson and Crick did not work alone to make this discovery, however, but instead relied on the work of many different people.

DNA was discovered in 1868 by Swiss physician Friedrich Miescher (1844–1895). He isolated a compound from the nucleus of cells, which he named nuclein. His work showed that the nucleic acids, which are long-chain polymers of nucleotides, were made up of sugar, phosphoric acid, and four nitrogen-containing bases. Later, it was found that the sugar in nucleic acid can be ribose or deoxyribose, which gives two forms: RNA and DNA. Only two years earlier, Gregor Mendel had finished his experiments with pea plants. Mendel showed that traits were inherited and passed from one generation to the next through genes. How genes carried this information was not known and the link between genes and DNA was not identified until 1944.

In 1944, American scientist Oswald Avery (1877–1955) worked with bacteria. He was not only able to transfer the ability to cause a disease to new, previously harmless bacteria, but also demonstrated that new bacteria could pass the disease on to its offspring. He did this by transferring nucleic acid from one bacterium to another. His work showed that nucleic acids, particularly DNA, were responsible for carrying the information in genes. The next question was to determine how DNA accomplished this.

In 1950, biochemist Erwin Chargaff (1905–2002) found that the arrangement of nitrogen bases in DNA varied widely, but the amount of certain bases always occurred in a one-to-one ratio. This meant that the pairs were always the same. Even though the molecule was only made up of four different bases, it somehow had to be able to store information. Everyone knew that the key to this puzzle was likely to be found in the structure of DNA.

Rosalind Franklin (1920–1958), a graduate in physical chemistry, had been working in Paris since 1947 on X-ray crystallographic methods and became an expert in the field of X-ray crystallography. In 1951, she went to work at the Medical Research Council Biophysics Research Unit at King's College in London to work on an X-ray picture of DNA that had been taken by a graduate student named Raymond Gosling. She was assigned the task of working on nucleic acids and X-ray diffraction of DNA along with another researcher named Maurice Wilkins.

Franklin found out that bundling super-thin strands of DNA and zapping them with a super fine X-ray beam resulted in two forms of hydration: the A form (easy to photograph), which is dry, and the B form (hard to photograph), which is wet. The B form photographs showed a fuzzy cross indicating that the DNA molecule was a helix. Franklin guessed that the backbone was on the outside and the bases were on the inside.

During a presentation at a seminar in November 1951, Franklin showed A and B form data. In the audience was American zoology graduate student James Watson, who was introduced to Franklin's work by colleague Maurice Wilkins. Watson became interested in molecular biology and the structure of DNA. He was working in England and became very excited about Franklin's talk. Watson returned to the University of Cambridge and got together with fellow student Francis Crick. They quickly assembled a model of DNA based on what Watson had remembered from the talk. Unfortunately, Watson did not take notes and he misremembered some important information. Their model turned out to be a triple helix. Watson and Crick then invited Franklin and Wilkins to Cambridge to see their model. Franklin and Wilkins immediately pointed out the flaws in the model and left. Watson and Crick had not only embarrassed themselves but also embarrassed their department. Their professor banned them from doing any further work on the DNA molecule.

In May 1952, Franklin finally was able to take a good photograph of the B form of DNA. The photograph showed a double helix. Franklin also continued working on the A form. She did not present or publish any of her data but instead, wanted to make sure she had all her facts assembled before she made any kind of announcement.

62 CELL THEORY

Figure 5.3 British biophysicist Rosalind Franklin took an X-ray diffraction image of DNA, called Photo 51. Scientist Maurice Wilkins passed the image and information on to James Watson, who, along with Francis Crick, discovered the structure of a DNA molecule, based on information provided in the image. The three men won the 1962 Nobel Prize in physiology or medicine for their findings; Franklin was not included because she died four years earlier and posthumous nominations for the Nobel Prize are forbidden.

At the same time, American scientist Linus Pauling (1901–1994) was also working on the structure of DNA. Pauling was a famous physical chemist with an interest in biological chemistry. In 1950, Pauling constructed the first model of a protein molecule. Pauling applied to attend a conference in England but was denied a visa. Had Pauling attended the conference, it is likely he would have met Franklin and they would have likely collaborated on their research.

Franklin became bogged down while trying to analyze her data of both the A and B forms of DNA. Her attempts at constructing a model did not pan out. She gave up on making a model for the B form to focus solely on the A form. At the same time, Watson and Crick were still trying to figure out the DNA model without the information that Franklin was finding and without the knowledge of their department.

In January 1953, Linus Pauling sent his son Peter, who was studying at Cambridge, a draft copy of his paper on DNA for comment. Watson knew Peter and managed to get a copy of Pauling's DNA paper. Watson went to see Franklin with a copy of the paper. Franklin dismissed the paper because she had written Pauling earlier for information, and Pauling had never responded. As Watson was leaving after his unsuccessful meeting with Franklin, he ran into Wilkins in the hall. He showed Wilkins the paper written by Pauling. Wilkins then gave Watson one of Franklin's photographs of the B form of DNA. A short time later, Crick was given a copy of a paper that contained all of Franklin's data.

Now that Watson and Crick had all Franklin's data, they saw that DNA was a multiple helix. Crick had worked on proteins and realized that Franklin's data implied an antiparallel double helix. He had just learned of Chargaff's findings about base pairs in the summer of 1952. At the same time, Franklin began working on the B form of DNA again. She already suspected that the A form of DNA was an antiparallel double helix but had not applied that idea to the B form.

Franklin was beginning to understand the implications of her data and began writing up her results. She completed the first draft of a paper indicating that her A form of DNA was a double helix. What she didn't know was that Watson and Crick had finished

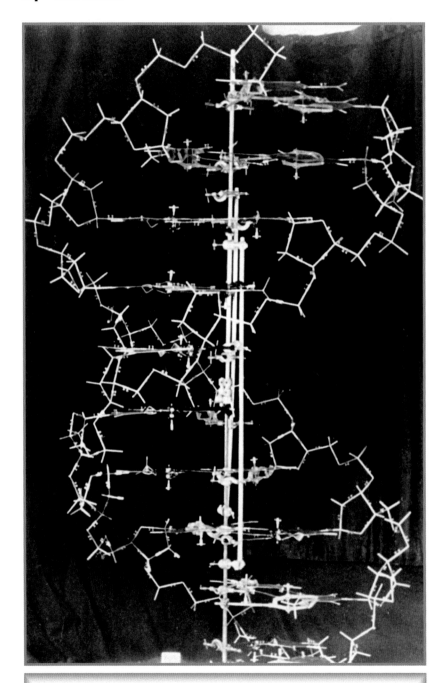

Figure 5.4 James Watson and Francis Crick created this DNA model to demonstrate their findings.

writing a paper for publication describing the double helix shape of DNA only one day earlier.

Watson and Crick's paper, "A Structure for Deoxyribose Nucleic Acid," was published in the journal *Nature* on April 25, 1953. The paper did not cite authorities or historical record. The first two sentences of the paper state, "We wish to suggest a structure for the salt of deoxyribose nucleic acid (DNA). This structure has novel features which are of considerable biological interest." That paper only stated a hypothesis of the shape and did not contain any experimental results. The only acknowledgement given to Franklin or others was the statement, "We have also been stimulated by a knowledge of the general nature of the unpublished results and ideas of Dr. M.H.F. Wilkins, Dr. R.E. Franklin, and their co-workers at King's College London."

In 1962, Watson, Crick, and Wilkins received the Nobel Prize in Physiology or Medicine for their work on the shape of the DNA molecule. Wilkins later won a Nobel Prize for his contributions to the structure of DNA. Franklin died before any of the Nobel Prizes were awarded and did not receive one because these prizes are only awarded to living scientists.

Human DNA

Human DNA is made up of about 3 billion base pairs. The DNA codes for an estimated 20,000 to 25,000 genes. The longest human chromosome, Chromosome 1, is made up of about 220 million base pairs. About 95% of the human **genome** is made up of junk DNA. Junk DNA does not serve a known function and is mostly leftover DNA from our evolutionary past. Even though the junk DNA does not seem to have a function, it may well be very important. In some cases, it seems that the junk DNA acts as spacers that keep genes separated. The junk DNA may also serve as a reservoir for potentially advantageous genes that may be needed in the future.

GENETIC MATERIAL AND THE CELL THEORY

One of the tenets of modern cell theory is that cells contain hereditary information (DNA), which is passed from cell to cell during cell division. This occurs because the DNA molecule is able to split apart. Because the structure of DNA is a double helix, or ladder-like, the DNA molecule is able to "unzip" to form two separate strands. Then, each strand serves as a template to reconstruct the other half of the helix.

DNA is made up of four nucleotide base units—thymine, guanine, adenine, and cytosine. The double-stranded DNA molecule is held together by chemical components called bases. Adenine (A) bonds with thymine (T); cytosine (C) bonds with guanine (G). The bonds make up the rungs of the ladder. Because nucleotides always bond to the same nucleotide, Chargaff's observation on the ratios of nucleotides holds true. The pairs of nucleotides contain instructions for groups of three base units, which are called codons. Groupings of these letters form the "code of life." There are about 2.9 billion base pairs in the human genome that are wound into 24 distinct bundles, or chromosomes. Written in the DNA are 30,000 to 40,000 genes that human cells use as templates to make proteins that build and maintain our bodies.

The Life of a Cell

According to the classical cell theory, the cell is the basic unit of life and structure. One of the tenets of the modern cell theory is that the cell is the structural and functional unit of life. Cells also give rise to other cells. These concepts are closely linked because cells must both carry on their functions and reproduce. These processes are very closely related.

Most cells reproduce by a process called binary fission, where they split into two parts. When a cell splits, it divides its organelles and its genetic material between the two new cells. Specialized sex cells undergo a different type of reproduction. This is because each sex cell requires half the genetic material because when sex cells combine, the resulting cell must have the same amount of genetic material as each parent cell.

An essential part of the cell's life is preparing for reproduction. Cells do this through a process called the cell cycle. This chapter discusses the cell cycle and how cells reproduce and will show how this process supports the cell theory.

THE CELL CYCLE

The **cell cycle** is a series of events that lead up to and include cell division. The cell cycle plays over and over for the life of a cell. This cycle has four stages, or phases. Each stage is designated by a code.

The stages are cell growth, production of RNA, and synthesis of protein (coded G_1); DNA synthesis and replication (S); preparation for reproduction (G_2); and mitosis (M).

Most of the time, cells are in the part of the cycle called the G_1 phase. At the beginning of this phase, DNA synthesis takes place. During this phase, the cellular activities operate at a high rate. The increased cell activity includes the synthesis of various enzymes that are required in the S phase, mainly those that are needed for DNA replication. The duration of the G_1 phase is highly variable, even among different cells of the same species. At the end of this phase, a cell starts to prepare for reproduction.

The S phase starts when DNA synthesis occurs. The DNA replicates to produce two complete copies of itself in each new chromosome. When DNA synthesis is complete, all of the chromosomes have been replicated into two chromatids, effectively doubling the amount of DNA in the cell. Activities such as protein synthesis take place at a very low rate during this phase. The duration of the S phase is relatively constant among cells of the same species.

After completing the S phase, the cell then enters the G_2 phase. This phase lasts until the cell starts mitosis. Most cellular activities are suspended except for the production of microtubules, which are required during the process of mitosis.

Mitosis is a very brief stage of the cycle but what happens here is very important. Mitosis is the process by which a cell duplicates the chromosomes to generate two identical daughter nuclei. This is followed by **cytokinesis,** which divides the nuclei, into two daughter cells. Each **daughter cell** contains roughly equal shares of the cellular components. Mitosis and cytokinesis together are the mitotic (M) phase of the division of the mother cell into two daughter cells, each of which is the genetic equivalent of the parent cell. Mitosis occurs exclusively in eukaryotic cells, but occurs in different ways in different organisms. Prokaryotic cells, which lack a nucleus, divide by a process called binary fission.

MITOSIS

The process of mitosis is complex and highly regulated. Mitosis follows a sequence of events or phases. Each phase is marked by the completion of one set of activities and the start of the next. The stages of mitosis are prophase, metaphase, anaphase and telophase. During mitosis, the chromosomes replicate into pairs. Each pair of chromosomes condense and fibers pull the sister chromatids apart. The cell then divides to produce two identical daughter cells.

The primary result of mitosis is the division of the parent cell's genome into two daughter cells. The genome is composed of a number of chromosomes. Each type of organism has its own genome of chromosomes. Within a species, the genes may vary, but each member of the species has the same number of chromosomes. Each chromosome is a complex of tightly coiled DNA that contain genetic information that is vital for proper cell function. When a cell undergoes mitosis, each resultant daughter cell should be genetically identical to the parent cell because the parent cell made a copy of each chromosome before mitosis.

Each new chromosome now contains two identical copies of itself that are called sister chromatids. Chromatids are attached together in a specialized region of the chromosome known as the centromere. The nuclear envelope that separates the DNA from the cytoplasm disassembles. The chromosomes align themselves in a line that spans the cell. Microtubules, which are thin fibers, splay out from opposite ends of the cell and shorten, pulling apart the sister chromatids of each chromosome. As a matter of convention, each sister chromatid is now considered a chromosome and are sometimes called sister chromosomes. The cell elongates and the corresponding sister chromosomes are pulled toward opposite ends. A new nuclear envelope forms around the separated sister chromosomes.

By the time the mitosis process is completed, cytokinesis is well underway. In animal cells, the cell pinches inward to separate the two developing nuclei. In plant cells, the daughter cells will construct a new dividing cell wall between themselves. Eventually, the mother cell will be split in half, giving rise to two daughter cells, each with a complete copy of the original chromosomes.

Prophase

Most of the time, the genetic material in the nucleus is in a loosely bundled coil that is called chromatin. At the beginning of prophase, the chromatin condenses or comes together into a structure called a chromosome. At this stage, the genetic material has already been duplicated earlier in the S phase, and each chromosome has two sister chromatids. The chromatids are bound together at the centromere. In most cells at this stage, the chromosomes are visible at high magnification with a light microscope.

Close to the nucleus are two centrosomes, which act as a coordinating center for the cell's microtubules. (The centrosomes are only present in animals. In plants, the microtubules form independently.) This marks the end of prophase and the beginning of metaphase.

Metaphase

During metaphase, spindle fibers align the chromosomes along the middle of the cell nucleus. This line is referred to as the metaphase plate. This organization helps to ensure that in the next phase, when the chromosomes are separated, each new nucleus will receive one copy of each chromosome. When this process is complete, the cell is ready for anaphase.

Anaphase

The paired chromosomes separate at the kinetochores and move to opposite sides of the cell. Motion results from a combination of

Figure 6.1 (opposite page) Mitosis is the process by which a cell reproduces itself and results in two identical daughter cells from a single parent cell. During interphase, the cell grows and the genetic material in the nucleus is duplicated. The cell then enters prophase in which the nuclear envelope breaks down, and the paired centrioles migrate to opposite sides of the cell while sending out fibers, forming the mitotic spindle. During metaphase, the chromosomes line up in the middle of the cell and fibers from both centrioles attach to each pair of chromosomes. Prometaphase is the stage during which the nuclear membrane begins to disintegrate. During anaphase, the daughter chromosomes are pulled by the spindle fibers to opposite sides of the cell and by late anaphase, as the daughter chromosomes near their destination, a cleavage furrow begins to form in the cell membrane indicating the beginning of cytokinesis. In the final stage, telophase, the cell membrane continues to constrict and eventually divides into daughter cells. As this is occurring, the nucleus is reestablished, and the daughter cells are once again in prophase.

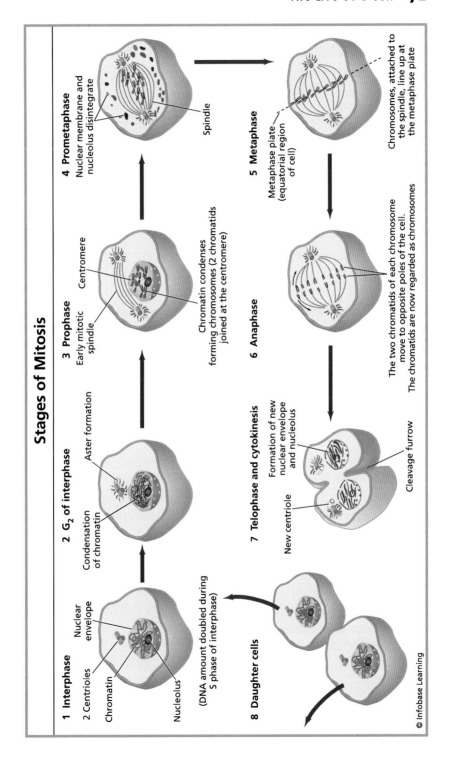

kinetochore movement along the spindle microtubules and through the physical interaction of polar microtubules. The sister chromatids are separated and are pulled apart by the kinetichore microtubules. They are then pulled toward the centromeres. The force that causes the centrosomes to move toward the ends of the cell is still unknown. One theory suggests that this movement may be caused by the rapid assembly and breakdown of microtubules. When the cell has finished separating the sister chromatids, the cell enters telophase.

Telophase

During telophase, chromatids arrive at opposite poles of the cell. New membranes form around each daughter nuclei. The chromosomes unfold into chromatin and are no longer visible under the light microscope. The spindle fibers disperse, and cytokinesis, or the partitioning of the cell, may also begin during this stage. Telophase is the phase where all the effects of mitosis are cleaned up and the new nuclei are formed. The cell is now ready to finish division with cytokinesis.

Cytokinesis

Cytokinesis is sometimes mistakenly thought to be the final part of telophase. However, cytokinesis is a separate process that begins at the same time as telophase. Cytokinesis is not even really a part of mitosis. Cytokinesis is a separate process that is necessary for cell division. The main event of cytokinesis is the physical dividing of the cell. In animal cells, a cleavage furrow pinches, isolating the separated nuclei into two new cells. In plants, the two daughter nuclei are isolated when a cell plate forms at the center of the elongated cell and develops into a cell wall. Each daughter cell has a complete identical copy of the genome of its parent cell. The end of cytokinesis marks the end of the mitosis phase.

The mitosis phase is important because each cell that is formed during this phase receives chromosomes that are alike in composition and equal in number to the chromosomes of the parent cell. The two daughter cells carry the exact same genetic information. The daughter cells function exactly like the parent cell and carry on the same functions. With the cell theory, this explains how all cells come from pre-existing cells.

MEIOSIS

Meiosis is a specialized type of cell division that results in four daughter cells that each have half of the genome of the parent cell. These types of cells take part in sexual reproduction by fusing with another cell that has half the genome of a different parent cell from a different organism. When two sex cells unite, this is called fertilization. The result is a mixing of genomes from two different parents. The new cell formed by this fusion has a unique genome that is made of two different genomes.

The process of meiosis is similar to mitosis. For the first part, the phases are the same as in mitosis: The cell goes through prophase, metaphase, anaphase, and telophase. This first part of the process is called meiosis I. Next, the two daughter cells go through meiosis II, which is similar to mitosis except that the chromatids do not

Cell Division

Cell division is a fascinating biological process that adds cells to organisms. The amount of time it takes for cell division to completely take place ranges widely, depending on the species. For example, some bacteria can complete cell division in 20 minutes. This means that a single bacterium can give rise to about 8 million bacteria in 24 hours. More complex organisms, such as humans, take about 80 minutes to complete cell division. This is a reflection not only of the complexity of the cells, but also the purpose of the cell division. Bacteria need to divide rapidly to reproduce and grow, while humans only need to replace cells that have reached the end of their lives.

74 CELL THEORY

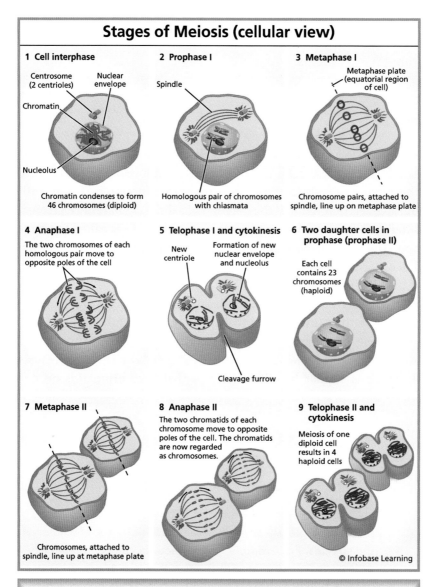

Figure 6.2 Meiosis is the process in which a cell divides and results in four sex cells with half the genetic content of the original parent cell.

replicate in meiosis II. The result is four daughter cells with half the genome of the parent cell. This allows for a mixing of genes that are passed on to the offspring.

The mixing of genes is good for passing on traits to offspring, but meiosis also has a mechanism for reshuffling genes. The reshuffling of genes occurs through two different processes: crossing over and random assortment. These two processes further shuffle the genes to provide a greater genetic variation of the offspring from either parent.

Crossing over occurs when non-sister chromatids swap sections. This is a reciprocal swap that keeps the same genes but will likely result in a swap of different genetic codes. Each chromatid will not have any gaps in their sequence resulting from crossing over. How this process takes place is still not fully understood, and it cannot be seen with a light microscope. Random assortment also comes into play during meiosis and is similar to shuffling a deck of cards. When a deck of cards is shuffled, the order of the deck is rearranged. In meiosis I, the orientation of chromatids at the metaphase plate is random. Therefore, although each cell produced by meiosis contains only one of each pair because cells have a number of different chromatids, the resulting combinations are a rearrangement of the parent cell.

CELL DIFFERENTIATION

Cell differentiation is the process by which a less specialized cell becomes a more specialized cell type. Cell differentiation is only found in multicellular organisms. Single-celled organisms only have one cell that must perform all of its required functions. Multicellular organisms are much more complex and require the division of labor that is afforded by cell differentiation.

Cell differentiation occurs many times during the development of a multicellular organism. As soon as the germ cells unite and form a zygote, the cells begin to specialize and form the various tissues and organs. Cell differentiation also takes place in adult stem cells. Stem cells divide and create fully differentiated daughter cells during tissue repair and during normal cell turnover. With a few exceptions, cell differentiation almost never involves a change in the DNA sequence. Even though all the different cells have very different physical characteristics, they still have the same genome.

Each specialized cell type in an organism expresses only a small number, or subset, of all the genes for that species' genome. The identity of each type of cell is determined by the particular pattern of genes expressed by the cell. Cell differentiation is the transition of a cell from one cell type by switching the pattern of genes expressed by the cell.

Cell differentiation plays an important part in the development of a multicelled organism. During development, cell differentiation is controlled by a regulatory gene. This gene determines which set of genes will be expressed by each cell. The result of this is a complete organism that has specialized cells that perform specialized functions.

CELL DEATH

Cells do not live forever. For example, they may die as a result of some type of trauma. Cells may also die through a process called **apoptosis**. Apoptosis is a form of programmed cell death in multicellular organisms. Apoptosis can also occur when a cell is damaged beyond repair, infected with a **virus**, or undergoing stress conditions such as starvation. Apoptosis involves a series of biochemical events that result in cell shrinkage, fragmentation of the nucleus, chromatin condensation, and DNA fragmentation.

Apoptosis is a part of the cell life cycle. It is important to the health of an organism that old cells be removed and replaced with new cells. Because cells continue dividing, old cells are removed to maintain the number of cells in an organism. In humans, between 50 billion and 70 billion cells die each day due to apoptosis.

Programmed cell death is an integral part of both plant and animal tissue development. The cells of an organ or tissue undergo extensive division and cell differentiation during development. This results in many extra cells that are removed by apoptosis. Unlike cellular death caused by injury, apoptosis results in cell shrinkage and fragmentation. This allows the cells to be easily removed and their components reused by the surrounding tissue.

Apoptosis is controlled by a number of different cell signals. These signals may originate within or outside of the cell. Signals from

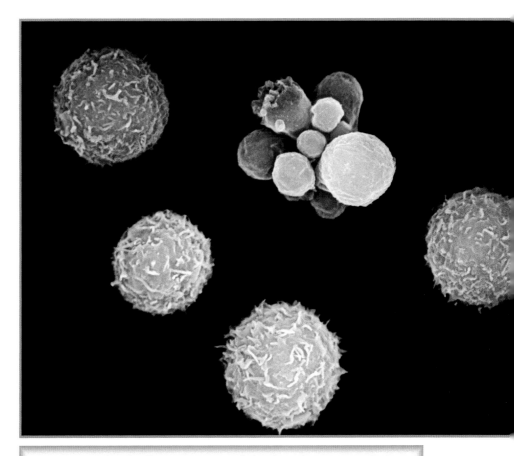

Figure 6.3 Blood cell apoptosis, or cell death, is shown, magnified 3,000 under a scanning electron microscope.

outside the cell include toxins, hormones, growth factors, or certain chemicals. Signals from within the cells may result from stress. This may ultimately result in apoptosis.

When a cell undergoes apoptosis, the cell releases enzymes into the cytoplasm. These enzymes begin breaking down the cell into smaller pieces. The nuclear membrane breaks down and the DNA fragments. Finally, the cell membrane breaks down and the cell breaks apart.

Applying and Questioning the Cell Theory

The cell theory has been around for almost 200 years. During this time, it is has been added to and modified slightly, but its basic tenets still apply to all living organisms. As scientists learn more about living organisms and search for extraterrestrial life, they keep returning to the cell theory as the definition of life.

So, what is life? The question sounds simple enough, but even scientists do not all agree on an answer. In general, they agree on a certain set of characteristics that all living things must have. A physical object that does not carry out all of a certain set of functions cannot be alive.

THE CHARACTERISTICS OF LIFE

Scientists have come up with a list of characteristics that define life. If something does not have all these characteristics, it cannot be alive. Most scientists agree with this idea. The list of characteristics varies and ranges from about 6 to 10 characteristics. What follows are the most commonly agreed upon characteristics of life.

- Living things are made up of cells.
- Living things need to take in energy.
- Living things get rid of waste.
- Living things grow and develop.
- Living things respond to their environment.
- Living things reproduce and pass their traits on to their offspring.
- Over time, living things evolve in response to their environment.

A scientist can use the list of characteristics to determine whether something is alive of not. Applying these characteristics is the first step in helping biologists study living things. As scientists continue to study living things and possibly even discover new forms of life outside of our planet, these basic characteristics may change.

Living organisms have traditionally been grouped into two broad categories: plants and animals. As scientists studied more about living organisms, they began finding organisms that did not fit exactly into one category or the other of these categories. This created the need to add more classifications. Currently, scientists use several different systems for classifying living organisms. These systems divide living organisms up into five or six kingdoms or three domains.

When classifying living organisms at the broadest level, scientists use either kingdoms or domains. The six kingdoms of life are Archaebacteria, Eubacteria, Protista, Fungi, Plantae, and Animalia. Some scientists prefer to group organisms into five kingdoms. This is accomplished by combining the Archaebacteria and Eubacteria into a single kingdom simply called Bacteria.

Other scientists prefer the three domain classification system. In this system, the three domains are a more general classification system which divides the six kingdoms up based on the organization of cell structures. The three domains are Archaea, Bacteria, and Eukarya. The Archaea Domain contains the Kingdom Archaeabacteria which are primitive prokaryots. The Bacteria Domain contains the Eubacteria Kingdom which are prokaryotes. The Eukarya Domain includes the Protista, Fungi, Plantae, and Animalia Kingdoms which are all eukaryotes.

ORIGINS OF LIFE

One of the tenets of the cell theory is that all cells arise from cells. This leaves open the question of where the first cell came from. The answer to this question is still openly debated among scientists. Many scientists believe that the first cell came from nonliving organic molecules on Earth. This theory is called **abiogenesis**. Abiogenesis has not been recreated in the laboratory. Under this theory, there are many different hypotheses on how exactly life came about. Some scientists even think that life on Earth may have actually arrived here on a comet. If that is true, the question of how the first life in

Viruses

Viruses are familiar to all of us. At one time or another, we have all had a virus, such as those that cause the common cold or the flu. Viruses are also responsible for many diseases and conditions such as measles, mumps, chicken pox, AIDS, and polio. Some viral infections are very mild while other may be life threatening.

A virus is little more than a short strand of DNA or RNA that is surrounded by a protein coating. A virus lacks any structures to utilize DNA or RNA, so it depends on the organelles of a host cell to carry out replication of the viral DNA or RNA. When a virus attacks a cell, it injects its DNA or RNA into the cell. The host cell is then taken over by the viral DNA or RNA and begins to replicate the genetic material and produces copies of the original virus. The copies are then released and infect more cells. This process continues and the infection spreads from cell to cell and eventually to a new host.

Viruses are generally not considered to be living organisms. They cannot reproduce without a living host cell. They are also not made up of one or more cells. Viruses have some of the characteristics of life, but they cannot perform functions on their own. As a result, they are not considered living organisms. Even under the cell theory, they do not fulfill all the basic tenets.

the universe started is a valid one. Many of these questions are hotly debated and are beyond the scope of this book. However, the subject is related to the cell theory.

According to one theory that scientists are fairly certain about, the building blocks of life were present in the oceans of the early Earth. These organic molecules made up what scientists call the "primordial soup." Over time, the molecules of the primordial soup became more and more complex until they engaged in complex chemical reactions. In some of these reactions, big molecules like proteins and nucleic acids actually started to copy themselves. Then, when the right kinds of self-copying molecules were trapped inside of the right kind of oil droplets, the first cells were formed. These first cells were able to reproduce themselves and were therefore subject to evolution by natural selection. They were also able to use the organic molecules floating around in the primordial soup as fuel and nutrients. To break down their fuel, they used anaerobic respiration. Thus, these first cells met the conditions for being living cells and they fit the tenets of the cell theory.

GENETIC ENGINEERING

Genetic engineering is a term applied to the field where the DNA is manipulated to create a new outcome. DNA manipulation takes place through gene isolation, gene insertion, gene transformation, gene splicing, and recombinant DNA. In all of these techniques, the gene expression in an organism is changed by forcing the expression of an existing gene or adding a gene from an organism of a different species. Using genetic engineering, scientists have made human insulin out of bacteria, created crops that produce their own insecticide toxins, and created experimental organisms for scientific research. Genetic engineering has also been used to make crops more uniform to increase production and simplify harvesting. As with any new technology, genetic engineering has proponents and opponents in regard to safety issues. Even as this debate continues though, genetically modified organisms, which are also called GMOs, are becoming more common.

Genetic engineering modifies the DNA in a cell. According to the cell theory, organisms contain hereditary information in DNA and pass it along to their offspring. How does this apply to modified DNA? Using genetic engineering to modify DNA is not really all that different from DNA changes that take place during natural evolution. Of course, adding genes from one organism into a different organism is much different from a change that takes place through evolution, but the end result is actually the same. Sometimes, genetically modified organisms are able to pass their new DNA on to offspring just as the cell theory predicts. Sometimes the passing of the genes to offspring may lead to an unintended consequence, but it does not invalidate this tenet of the cell theory.

CANCER RESEARCH

Cancer is a term used to describe a class of diseases in which abnormal cells divide without control and are able to invade other tissues. Cancer cells may also spread to other parts of the body through the blood and lymph systems. Cancer cells undergo uncontrolled growth and reproduction because of a disrupted cell cycle. Cancer cells are also not controlled by normal apoptosis. As a result, cancers usually grow quickly and spread through and into other tissues. Nearly all cancers are caused by abnormalities in the genetic information. These abnormalities lead to their excessive growth.

Cancer affects all organisms, including humans. It also has many different causes. Sometimes, the genetic material in cells is damaged through exposure to chemicals or by viruses. Other times, the damage is caused by exposure to radiation or ultraviolet rays. Sometimes, cancer may be caused by random mutations. Some people seem more susceptible to cancer than others. Some people seem to have an easier time fighting cancer. Scientists are working to find ways to prevent and treat all types of cancer.

Cancer is not just one disease but many diseases. There are more than 100 different types of cancer. Most cancers are named for the organ or type of cell in which they start. For example, cancer that begins in the lungs is called lung cancer and cancer that begins in the colon is called colon cancer.

All cancers begin in cells, the body's basic unit of life. To understand cancer, it's helpful to know what happens when normal cells become cancer cells. Remember that normal cells grow and divide in a controlled way to produce more cells. When cells become old or damaged, they die and are replaced with new cells. When the

DNA of a cell is damaged or changed, the cell produces mutations that affect normal cell growth and division. When this happens, cells do not die when they should and new cells that the body does not need form. The extra cells may form a mass of tissue called a tumor. Not all tumors are cancerous. These noncancerous or benign tumors do not spread to other parts of the body and are often easily removed. However, some tumors do continue to grow and become cancers. Other cancers, such as leukemia or blood cancer, do not start as tumors.

Cancers do not violate any of the tenets of the cell theory. Their growth simply does not follow the normal guidelines for cell growth. Scientists are looking into normal cells to try and find ways to regulate cancerous cells to prevent them from becoming cancers. Before the development of the cell theory, scientists were not sure how cancer formed.

Cancer was poorly understood until the anatomy and functions of the human body were better understood. Advances in the understanding of blood circulation, the function of the lymphatic system, along with the development of the cell theory, gave scientists insight into cancer. Once this knowledge was in place, theories of cancer developed. By the mid 1800s, this knowledge led to the description of leukemia and lymphatic cancer. In 1914, the theory of cancer was established by Theodor Boveri. He proposed that all cancers were caused by chromosome abnormalities within living cells, or by agents or events that produce these abnormalities. This theory was reinforced by the discovery that tumors could be caused by viruses or chemicals in the environment, and it is now generally accepted.

STEM CELL RESEARCH

Stem cells are a type of cell that is found in multicellular organisms. Stem cells are characterized by their ability to renew themselves through mitosis and differentiate into specialized cell types. Stem cells have two important characteristics that distinguish them from other types of cells. First, stem cells start out as unspecialized cells that renew themselves for long periods through cell division. Second, under certain conditions, they can be induced to become cells with specialized functions.

84 CELL THEORY

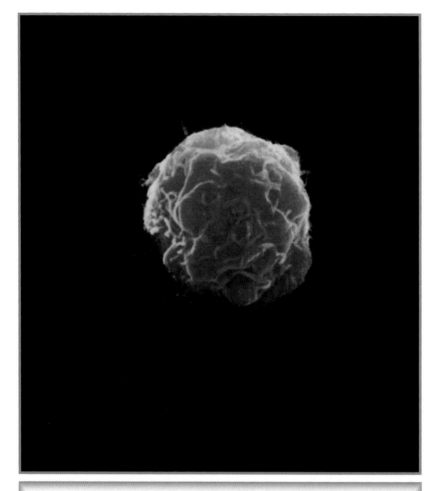

Figure 7.1 Breast cancer is a malignant tumor—group of cancerous cells—that starts in cells in the breast. This scanning electron micrograph shows a breast cancer cell.

Stem cells are important to living organisms for many reasons. Stem cells are classified into two different categories—embryonic and adult stem cells. In a developing embryo, which is called a blastocyst, embryonic stem cells give rise to the multiple specialized cell types that make up the heart, lung, skin, and other tissues. In some adult tissues, such as bone marrow, muscle, and brain, adult stem cells generate replacements for cells that are lost through normal wear and tear, injury, or disease. Because of these abilities, scientists have started studying stem cells to find ways to help the body repair itself when injured or damaged.

Stem cell research supports many of the tenets of the cell theory. The cell is the basic structural unit of organisms. Multicellular organisms have cells that are specialized. Because all cells have basically the same composition, cells must express different genes in their DNA to perform specialized functions. Stem cells have all of this information in their DNA and different parts can be expressed or turned on to produce a certain type of cell. Researchers are particularly interested in this property because it can be use to repair damaged body parts. The more scientists learn about stem cells, the more the cell theory is reinforced.

NANOBACTERIA

Nanobacteria are much smaller than regular bacteria. Nanobacteria are minute, about one-tenth the size of bacteria. They can be highly variable in appearance, but usually appear as spheres or rods. The only constant factor seems to be their coat of apatite, which is formed from soluble calcium and phosphorus compounds in their environment. Their existence was suspected in the early 1980s. At that time, because they were smaller than the smallest known bacteria, they were called ultramicrobacteria. They were also not readily visible under a microscope, so little was known about them. It was not until 1989 that Robert L. Folk, a geologist, isolated nanobacteria from a rock sample from Italy. The sample was of a rock called travertine, which is formed from mineral-laden water from a hot spring. In 1992, Folk presented a paper at the Geological Society of America Conference. In his paper, he presented his findings and further stated that nanobacteria (sometimes spelled with two *n*s), or nanoparticles or nanobes, were the responsible agent for causing the precipitation of minerals and crystals from crystal solutions. His radical idea was not immediately accepted.

Additional work by Folk and others showed that these nanobacteria were actually quite common and widespread. Other scientists confirmed Folk's theory that minerals and crystals were formed with the aid of nanobacteria. As scientists began to accept this theory, a new discovery was made when a meteorite from Mars was

86 CELL THEORY

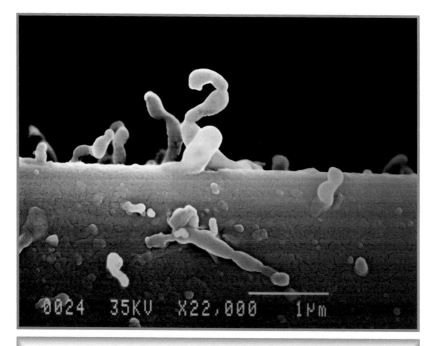

Figure 7.2 This colored scanning electron micrograph (magnified 12,000 times) shows nanobacteria on sandstone. Thought to be the smallest self-replicating organisms, they range from 20 to 150 nanometers in size, smaller than any other bacteria. They were found in 200 million year old sandstones taken from about 2 to 3 miles (3 to 5 km) below the seabed. When the rocks were opened, the nanobacteria multiplied and grew in the laboratory. Tests show they contain DNA.

found in Antarctica containing fossils that looked like nanobacteria. However, whether or not these are actually fossils of nanobacteria is still being debated by scientists. In 1998, Finnish researcher Olavi Kajander and Turkish researcher Neva Ciftcioglu, working at the University of Kuopio in Finland, grew what they called nanobacteria. Their work implicated nanobacteria in arterial plaque development and kidney stone formation in humans. It seemed that when scientists started looking for nanobacteria, they found them everywhere.

By 2008, some researchers reached a new conclusion. They believe that nanobacteria are not really living organisms after all. They believe that they are some type of self-replicating mineral crystal that has some but not all of the characteristics of life. With this new direction of research, some scientists are leaning toward considering nanobacteria as nonliving yet still a possible cause for some medical problems and some geological phenomenon. More work in the coming years will help determine whether nanobacteria are living or nonliving as well as being biological or geological.

So, how do nanobacteria relate to the cell theory? Nanobacteria are much smaller than bacteria. Some believe that they are too small to have DNA, RNA, or plasmid (circular bacterial DNA). Yet nanobacteria seem to reproduce and exist on their own. Unlike viruses, nanobacteria do not need a host to perform some of their life functions. They seem to have a cell wall even though they have a mineral coating. However, because they are so small, they are difficult to observe. Scientists still debate their existence.

The examples described here show how the cell theory is applied to new situations. As scientists continue researching and making new discoveries, the cell theory will continue to be tested. Testing is how scientists validate and expand on a theory.

Nanobacteria and Caves

Nanobacteria seem to be common in more extreme environments. One of these environments is underground, in caves. Caves generally have a constant temperature, but they lack sunlight. As a result, they tend to be low energy environments because there are no plants to convert sunlight into food. As a result, cave-dwelling organisms tend to have a long life spans and low metabolisms. Nanobacteria found in caves are also suspected of being responsible for the creation of spectacular caves formations, ranging from tiny, delicate crystals to huge stalactites and stalagmites. Nanobacteria may also be responsible for the flowstone and drapery formations found in some caves. Microbiologists are studying nanobacteria and are hoping to find new pharmaceutical products associated with these organisms.

The Cell Theory and Modern Biology

Biology literally means "the study of life." Biology examines the structure, function, growth, origin, evolution, and distribution of living things. Biologists classify and describe organisms, determine their functions, study how species come into existence, and investigate how organisms interact with each other and with the natural environment. Biology is guided by four unifying themes that form the foundation of modern biology—cell theory, evolution, genetics, and homeostasis.

THEMES OF MODERN BIOLOGY

The cell theory is one of the themes of modern biology. The cell theory identifies the cell as the basic structural and functional unit of all known living organisms. The cell theory dictates that cells come from pre-existing cells. The cell is the smallest unit of an organism that is classified as living. The cell is the building block of life.

Because the study of biology is the study of life, the cell is a central subject of interest in the field. Biologists strive to understand how the cell functions to discover clues about how life functions. As the basic unit of life, the cell serves as a model for not only unicellular organisms but also multicellular organisms. In multicellular organisms, cells become specialized. Even though they have the same genetic material as the other cells in the organism, specialized cells perform specialized functions. This division of labor within an organism is also similar to the way different organisms interact within their environment. Essentially, all organisms depend on other organisms. This complicated web of interaction is explored in the study of biology and its related scientific fields.

As mentioned, the study of biology is guided by four unifying themes. The cell theory is one of these themes. We have explored the underlying concept of the cell theory from its development, through its expansion, and finally its application in biology. The cell theory has been in place for almost 200 years and, even though it has been refined, its basic tenets still stand. As scientists learn more about life on Earth and search for life in our solar system and beyond, the cell theory will be a guiding principle of research. The cell theory is also interwoven through the three other unifying themes in biology.

Another unifying theme of biology is evolution. Evolution is the process of change in the inherited traits of a population of organisms from one generation to the next. The inherited traits are passed from one generation to the next through the genes. This passing of genes forms the basis for evolution. Some genes become mutated and produce new or altered traits. If these traits are passed on to the offspring, the new traits may contribute to the evolution of the species. New genes may also be introduced into a population by mixing genes with other isolated populations or between different species. The new genes may also be produced by genetic recombination during sexual reproduction. Evolution only occurs when these heritable differences become more common or rare in a population.

Evolution takes place by two mechanisms: natural selection and genetic drift. Natural selection occurs when an organism successfully adapts to a particular environment and then passes the genes for that adaptation to its offspring. Over many generations of this process, the traits for better adaptation make it possible for organisms to successfully survive in a particular environment. Over more time, populations may become specialized to the point where they are considered a new species. Genetic drift is the other mechanism for evolution. Genetic drift occurs naturally in populations that are isolated from the rest of the world, such as species that live on a far

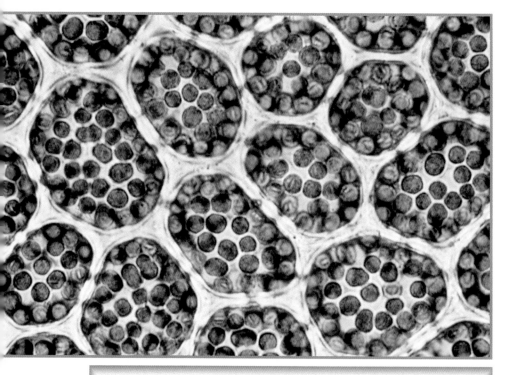

Figure 8.1 Plant cells (*magnified here*) are like factories. They take in raw materials (carbon dioxide and water) and they produce glucose and oxygen using energy captured from the sun. Plants provide the food needed by animals.

away island. Isolated populations do not interbreed with other populations, so they do not exchange genetic information. Over time, changes in expressed genes and traits in these isolated populations may cause enough change or drift in the genomes to consider each population a different species.

Another unifying theme in biology is genetics. Genetics is the study of heredity. Heredity is the passing of genes from parents to offspring. In bacteria, reproduction takes place through binary fission where the bacteria simply divide into two new bacteria. Each new bacterium has identical genes. These bacteria continue to reproduce through binary fission and continue making identical genetic copies. Sometimes a mutation develops and a new genetic makeup is started. Other organisms reproduce through sexual reproduction. Sexual

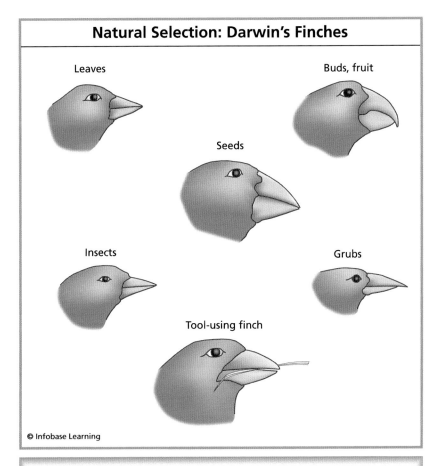

Figure 8.2 A famous example of evolution through natural selection is the finches of the Galapagos Islands, as noted by naturalist Charles Darwin. Over time, the population of finches on these isolated islands in the Pacific Ocean evolved into separate species—with different beak sizes and shapes based on the food they ate.

reproduction involves the uniting of a sex cell from two different parents. The resulting offspring has half of its genetic material from each parent. The result is an offspring with a different genetic makeup from each parent. This quickly creates genetic variation within a population. Genetics is the study of how these traits are passed from parent to offspring and how genes are expressed as traits.

The final unifying theme of biology is homeostasis. Homeostasis is the maintenance of a suitable environment within an organism and its cells. In a single-celled organism, the cell has mechanisms that regulate the amount of water and chemicals within the cell, regardless of the external environment. A cell performs this function by having a membrane that separates its interior from the exterior environment. The membrane is permeable, so water may pass through the membrane in each direction. The cell has tiny pumps that move excess water out of the cell or pull water into the cell to maintain a balance. If these pumps are overwhelmed, the cell will either swell up and explode or wither away. Maintaining the internal conditions of the cell is a critical function and a task that requires much energy. Multicellular organisms have additional requirements for homeostasis. Not only does each individual cell need to maintain homeostasis but the organism as a whole must also maintain homeostasis. The requirements for homeostasis vary from organism to organism.

Each of these unifying themes in biology is interwoven. Each theme depends on the other themes to deal with the problems that each organism faces and to find solutions. Each of these themes apply at the cellular level, the organism level, and the population level. The field of biology simply cannot be described or grow in knowledge without seeing all of these themes as an interwoven, unifying concept.

CELL THEORY AND EVOLUTION

The cell theory and evolution are very closely linked together. One of the big questions faced by evolution is the origin of life. As discussed earlier, part of the cell theory dictates that all cells arise from other cells. The unanswered question is: Where did the first cell come from? This is similar to the question, which came first, the chicken or the egg? The question is difficult to answer. No one has created or observed the formation of a living cell from inanimate chemicals. When—and this may not happen for a very long time— the origin of life is determined, it will certainly call for a modification of the current cell theory. It is likely that the modification will only affect the first tenet of the cell theory, but it is difficult to predict what will be discovered in the future. Scientists may discover the need for a new cell theory that will completely replace the current cell theory.

CELL THEORY IN MODERN RESEARCH

Modern research into cells and their functions has reinforced the cell theory, but it also has expanded on the original cell theory. The only part of the original theory that has been replaced has been spontaneous generation. This development took place after Rudolf Virchow discovered that cells were actually derived from other cells. As previously discussed, the actual origin of the first cell is unknown and may present a challenge to the cell theory when its origin is determined.

Modern research continues to discover new information about cells and how they function. Each new discovery so far has not only added to our knowledge but has reinforced the modern cell theory. Future advances in cell study may add more tenets to the cell theory. Future studies will certainly increase our understanding of how the cell theory relates as one of the unifying themes of biology.

WITHSTANDING THE TEST OF TIME

The cell theory has stood for almost 200 years and has only been expanded and slightly modified. When you read about the cell theory, you are likely struck by its simplicity. Today, we almost take this simplicity for granted because of how it makes sense. However, our knowledge of the cell theory is based on knowledge that we have accumulated over the years. Before you had ever heard of the cell theory, you had already learned the basics. When you heard about the cell theory for the first time, you probably just accepted it. Remember though that the scientists who first developed the theory did not have the advantage of your knowledge. Those scientists were breaking new ground and spent their lifetimes developing and proving the theory. The cell theory results from the work of many great scientific minds.

As time advances, the cell theory has been, and will continue to be, tested. Scientists are hard at work in many fields related to it. It is possible that in the future, new information will be discovered that either further validates or invalidates the cell theory. It is possible that as scientists search for life in our solar system and beyond, new life forms may be found that do not conform to the cell theory. Should that happen, a new theory will be found to replace it.

Glossary

abiogenesis The theory that states that life originated from non-living organic molecules

apoptosis Cell death

bacteria Single-celled organisms that reproduce through binary fission

cell The smallest unit of life

cell cycle The cycle a cell follows as it grows and divides

cell differentiation Process by which cells become specialized

chromosome Tightly coiled DNA

cytokinesis The dividing of cell material during cell division

cytoplasm The fluid matrix of a cell

daughter cell Cells that result from cell division

deoxyribonucleic acid (DNA) A helical shaped, long-chain organic molecule that contains genetic information

electron microscopes An instrument that uses a beam of electrons to magnify an object

enzyme A protein that acts as a catalyst to facilitate a chemical reaction

eukaryotes Advanced cells with organelles bound to its membrane and a nucleus

genetic material Chromosomes that are found in the nucleus of a cell; they may be passed to offspring.

genome The sum of all genetic material in an organism

germ cells Specialized cells that take part in sexual reproduction

meiosis   Specialized type of cell division that results in four daughter cells, with each cell having half the genetic material of the parent cell

microscopes   An optical instrument that magnifies an object

mitosis   The process by which a cell duplicates and forms two genetically identical daughter cells

neurons   Nerve cells that conduct electrical impulses

nucleic acid   A group of organic molecules that are made up of long chains of nucleotides

nucleus   The control center of a cell that contains the DNA

organelles   The structures in a cell that carry out specific functions

plasma membrane   A phospholipid sac that surrounds a cell and separates it from its surrounding environment

prokaryotes   Primitive cells that lack a nucleus

ribonucleic acid (RNA)   A long chain organic molecule that contains instructions for making proteins in a cell

virus   A disease causing agent made up of a short strand of DNA or RNA surrounded by a protein coat

Bibliography

Alberts, B., D. Bray, K. Hopkins, A. Johnson, J. Lewis, M. Raff, K. Roberts, and P. Walter. *Essential Cell Biology*. London: Garland Science/Taylor & Francis Group, 2003

Becker, W.M., L.J. Kleinsmith, J. Hardin, and G.P. Bertoni. *The World of the Cell* (7th ed.). San Francisco: Benjamin Cummings, 2008.

Coleman, W. *Biology in the Nineteenth Century: Problems of Form, Function and Transformation*. London: Cambridge University Press, 1978.

Cooper, G.M. *The Cell: A Molecular Approach, 2nd ed.,* Washington, D.C.: ASM Press, 2000.

Gould, S.J. *The Structure of Evolutionary Theory*. Cambridge, Mass.: Belknap Press, 2002.

Johnson, A.B., A. Lewis, L.J. Raff, and R.K. Walter. *Molecular Biology of the Cell, 4th ed.* London: 2002.

Lodish, H., A. Berk, P. Matsudaira, C.A. Kaiser, M. Krieger, M.P. Scott, S.L. Zipurksy, and J. Darnell. *Molecular Cell Biology, 5th ed.,* New York: WH Freeman, NY, 2004.

Magner, L.N. *A History of the Life Sciences, Third Edition*. Boca Raton, Fla.: CRC, 2002

Mayr, Ernst. The *Growth of Biological Thought: Diversity, Evolution, and Inheritance*. Cambridge, Mass.: Belknap Press, 1985.

Mazzarello, P.A. "Unifying Concept: The History Of Cell Theory," *Nature Cell Biology* 1, (1999) E13–E15.

McElheny, V.K. *Watson and DNA: Making a Scientific Revolution*. New York: Basic Books, 2004.

Moore, J.A. *Science as a Way of Knowing: The Foundations of Modern Biology*. Cambridge, Mass.: University Press, 1999.

Serafini, A. *The Epic History of Biology*. New York: Basic Books, 2001.

Watson, J.D. *A Passion for DNA: Genes, Genomes, and Society*. Woodbury, NY: Cold Spring Harbor Laboratory Press, 2000.

Further Resources

Cobb, A.B. *Scientifically Engineered Foods: The Debate over What's on Your Plate.* New York: Rosen Publishing Group, 2003.

Liebes, S. *A Walk Through Time: From Stardust to Us, the Evolution of Life on Earth.* New York: Wiley, 1998.

Margulis, L. and D. Sagan. *What Is Life?* Los Angeles: University of California Press, 2000.

Taylor, M.R. *Dark Life: Martian Nanobacteria, Rock-Eating Cave Bugs, and Other Extreme Organisms of Inner Earth and Outer Space.* New York: Scribners, 1999.

Thomas, L. *The Lives of a Cell: Notes of a Biology Watcher.* New York: Penguin, 1995.

Watson, J. *The Double Helix: A Personal Account of the Discovery of the Structure of DNA.*, NAL, 1991.

Web Sites

Action Bioscience
http://www.actionbioscience.org
> This site provides peer-reviewed articles by scientists, science educators, and science students.

Cells Alive!
http://www.cellsalive.com/
> Find interactive cell models, puzzles, and a cell cam that takes viewers through the stages of cell division among cancer cells.

The Cell Biology Project
http://www.biology.arizona.edu/cell_bio/cell_bio.html
> This site a clearinghouse of information about cells.

The Virtual Cell
http://www.ibiblio.org/virtualcell/tour/cell/cell.htm
> Take an animated tour through a cell to see the structures and organelles and learn about their functions.

Picture Credits

Page

14: © Infobase Learning
17: © Infobase Learning
19: © Infobase Learning
23: © Infobase Learning
26: © Infobase Learning
29: © Inga Spence/Visuals Unlimited, Inc.
32: © Infobase Learning
35: National Library of Medicine/U.S. National Institutes of Health
37: © Adam Hart-Davis/Photo Researchers, Inc.
41: © Prisma Archivo/Alamy
43: © Infobase Learning
44: © SSPL via Getty Images
49: © Infobase Learning
52: © Dennis Kunkel Microscopy Inc./Visuals Unlimited, Inc
57: © Professors Pietro M. Motta & Sayoko Makabe/Photo Researchers, Inc.
58: National Library of Medicine/U.S. National Institutes of Health
62: Courtesy of Jenifer Glynn/National Library of Medicine/U.S. National Institutes of Health
64: © James D. Watson Collection/Cold Springs Harbor Laboratory Archives
71: © Infobase Learning
74: © Infobase Learning
77: © Dr. Gopal Murti/Visuals Unlimited, Inc.
84: © Phototake Inc./Alamy
86: © Dr. Philippa Uwins, Whistler Research Pty./Photo Researchers, Inc.
90: © Michael Eichelberger/Visuals Unlimited, Inc.
91: © Infobase Learning

Index

A

Abbe, Ernest, 25–27
abiogenesis, 80–81
adenine, 66
adult stem cells, 84
algae, 20
alleles, 59
amino acids, 12
anaerobic respiration, 81
anaphase, 69, 70–72, 73
anchor escapements, 38
animalcules, 35
Animalia kingdom, 79
apochromatic lenses, 27
apoptosis, 76–77
Archaea domain, 79
Archaebacteria kingdom, 79
assortment, random, 75
atoms, in cells, 11
ATP (adenosine triphosphate), 14
Avery, Oswald, 60
axons, 48, 49, 52

B

bacteria, 8
Bacteria domain, 79
Bacteria kingdom, 79
balance springs, 38
binary fission, 39, 67, 90. *See also* Cell division
Binnig, Gerd, 30, 31
biology, defined, 88
Boveri, Theodor, 83
Boyle, Robert, 36
brain, anatomy of, 47
Brown, Robert, 40

C

Cajal, Ramón y, 53, 54
cambium, 39
cancer, 82–83
carbohydrates, 11–12
carbon, organic molecules and, 11
carotenoids, 20–21
caves, nanobacteria and, 87
cell body, neuron structure and, 48, 52
cell cycle, 67–68. *See also* Meiosis; Mitosis
cell death, 76–77
cell differentiation, 75–76
cell division, 14, 39–40
cell walls, 17, 20
cellulose, 11, 17, 20
centromeres, 69

centrosomes, 14, 69–70
Chargaff, Erwin, 60, 66
chlorophyll, 17, 20–21
chloroplasts, 17, 18–19, 20
chromatids, 69–72
chromatin, 69–70
chromosomes. *See also* Meiosis; Mitosis
 cancers and, 83
 function and location of, 14–15
 heredity and, 55–59
 human DNA and, 65
Ciftcioglu, Neva, 86
classical cell theory, 7–9
codons, 66
combustion, 38
composition, 30–31
compound microscopes, 22–24, 27
cork, 25, 36
Crick, Francis, 60, 61–65
crossing over, 75
crystallization, 85–87. *See also* Free cell formation
crystallographic information, 31
cyanobacteria, 18–19

cytokinesis, 68, 69, 72–73. *See also* Mitosis
cytology, 39
cytoplasm, 13, 14, 15, 20, 69
cytosine, 66
cytoskeleton, 16

D

Darwin, Charles, 7, 55–56, 91
daughter cells, 68
de Graal, Regnier, 36
Deiters, Karl, 52–53
dendrites, 48, 49, 52
diaphragms, 27
differentiation, cellular, 75–76
DNA (deoxyribonucleic acid)
  discovery of, 60–66
  endosymbiotic theory and, 18
  genetic engineering and, 81–82
  location and function of, 13–15
  PCR and, 33
DNA synthesis and replication (S stage), 68
domains, 79
dominant traits, 59
Drosophila, 59
Dumortier, Barthelemy, 39
Dutrochet, René Joachim Henri, 40

E

eggs, 22
electron microscopes, 28–32
embedding, 31–32
embryology, 38–39
embryonic stem cells, 84
endoplasmic reticulum (ER), 14, 15
endosymbiotic theory, 18–19
energy, 11–12, 15–16
enzymes, 12, 16
epoxy, embedding in, 31–32
ER (endoplasmic reticulum), 14, 15
Eubacteria kingdom, 79
Eukarya domain, 79
eukaryotes, 14, 18–19, 20, 68. *See also* Plant cells
evolution, 56, 89–90, 92
eyeglasses, 24
eyepiece, 27

F

facts, theories vs., 10–11
fats, 12
fertilization, 73
fission, binary, 39, 67, 90. *See also* Cell division
Flemming, Walther, 56
flowstone, 87
Folk, Robert L., 85
Franklin, Rosalind, 61–65
free cell formation (spontaneous generation), 8, 39, 42, 45

Fungi kingdom, 79

G

G_1 stage of cell cycle, 68
G_2 stage of cell cycle, 68
Galen, 47
Galileo Galilei, 24
gemmules, 56
gene shuffling, 75
genetic drift, 89–90
genetic engineering, 81–82
genetics, study of, 90–91
genome, human, 65
germ cells, 56
glia, 50
glucose, role of, 11–12
glycogen, 11
Golgi, Camillo, 53
Golgi apparatus, 14, 15
Gosling, Raymond, 61
grasshoppers, 59
guanine, 66

H

helix structure of DNA, 61–65
hemicellulose, 17
heredity, 56–60, 66, 89–91
His, Wilhelm, 53
homeostasis, 91–92
Hooke, Robert, 7, 24–25, 26, 36–38
Hooke's law, 38
Huygens, Christian, 36
hybrids, 59

I

illumination sources, 27
immersion lenses, 27, 28
inheritance, 56–60, 66, 89–91
intermediate filaments, 16
intermediate neurons (interneurons), 49
International System of Units (SI), 16
iris diaphragm, 38

J

Janssen, Hans, 24
Janssen, Zaccharias, 24
junk DNA, 65

K

Kajander, Olavi, 86
kinetochores, 70–72
kingdoms, 79
Knoll, Max, 28, 30

L

Lamarck, Jean-Baptiste, 55
leukemia, 83
life, characteristics and origins of, 78–81
lignin, 17
lipid bilayer, 13
lipids, 11, 12, 13, 15
Lippershey, Hans, 24
lysosomes, 14, 16

M

M stage of cell cycle (meiosis), 73–75
Margulis, Lynn, 18
meiosis, 59, 73–75
meiosis I, 73
meiosis II, 73–74
membranes, 12, 18, 39. *See also Specific membranes*
Mendel, Johann Gregor, 56–59, 60
metabolism, overview of, 13
metaphase, 69, 70, 73
metaphase plate, 70
Meyen, Franz Julius Ferdinand, 39
microfilaments, 16
Micrographia (Hooke), 25, 38
microns, defined, 16
microscopes
  electron, 28–32
  history and development of, 24–25, 26, 36
  modern optical, 25–28
  overview of, 22–24
microtubules, 14–16, 70–72
Miescher, Friedrich, 60
minerals, crystallization of, 86
Mirbel, Charles-François Brisseau de, 39
mitochondria, 14–16, 18–19
mitosis, 56, 68–73
Moldenhawer, Johan Jacob Paul, 39
Morgan, Thomas Hunt, 59
morphology, 30
motor neurons, 49
multicellular organisms, defined, 22
myelin sheath, 48–50

N

nanobacteria, 85–87
natural selection, 89
nervous system, overview of, 47–54
neuroglia, 49
neuron doctrine, 53–54
neurons, 48–54
neurosecretory neurons, 49
neurotransmitters, 48
Newton, Isaac, 36, 37
Nobel Prizes, 31, 62, 65
nuclear membrane, 15
nucleic acids, 11, 12–13
nuclein, 60
nucleoli, 15
nucleotides, 12–13
nucleus, 13, 15, 40

O

objective lenses, 27, 28
ocular, 27
oil-immersion lenses, 28
oils, 12
organelles, 13, 14, 15, 20, 32. *See also Specific organelles*
organic molecules, 11
osmosis, 40
osmotic pressure, 20
ostrich eggs, 22

P

pangenesis, 56, 57
parenchymatous tissue, 39
Pasteur, Louis, 43–45
pathology, 42
Pauling, Linus, 63
Pauling, Peter, 63
PCR (polymerase chain reaction), 32, 33
pea plants, 56–59, 60
pectin, 17
peer review, scientific theory and, 10
peroxisomes, 14
phlegms, 47
phospholipids, 12, 13
Photo 51, 62
photosynthesis, 11, 17
Phytotomie (Meyen), 39
plan apochromatic lenses, 27
plant cells, 17, 20–21, 39–40
Plantae kingdom, 79
plasma membrane, 13–14
plastids, endosymbiotic theory and, 18
polymerase chain reaction (PCR), 32, 33
polysaccharides, 11–12, 20
primordial soup, 81
programmed cell death, 76–77
prokaryotes, 17, 18–19, 20, 68
prometaphase, 70
prophase, 69–70, 73
protein synthesis. *See* G_1 stage of cell cycle
proteins, 11, 12, 14, 15
proteobacteria, 18–19
Protista kingdom, 79
protists, 8

R

random assortment, 75
recessive traits, 59
Reddi, Francesco, 42–43
reproduction, 8–9
respirators, 38
review, peer, scientific theory and, 10
ribosomes, 14, 15
RNA (ribonucleic acid), 13, 33, 60, 80
Rohrer, Heinrich, 30, 31
Ruska, Ernst, 28–30, 31

S

S stage of cell cycle, 68
scanning electron microscopes (SEM), 30
scanning objective lens, 28
scanning tunneling electron microscopes (STEM), 30, 31
Schimper, Andreas, 18
Schleiden, Matthias Jakob, 7–8, 40–42
Schott, Otto, 25
Schwann, Theodor, 7–8, 40–42
scientific theory, overview of, 9–11
SEM (scanning electron microscopes), 30
sensory neurons, 49
sex determination, 59
sexual reproduction, 56, 90–91
shuffling of genes, 75
SI system, 16
sister chromatids, 69–72
solvent, water as energy and, 11
somatic cells, 56
sperm, 56
spindle apparatus, 14
spindle fibers, 70
spontaneous generation (free cell formation), 8, 39, 42, 45
stage, 27
stalactites and stalagmites, 87
Stand I microscope, 25
starch, 11
STEM (scanning tunneling electron microscopes), 30, 31
stem cells, 75–76, 83–85
stomata, 39
Sutton, Walter, 59

T

telophase, 69, 73
TEM (transmission electron microscopes), 28–30
theories, facts vs., 10–11
thymine, 66
timing of cell division, 73

topography, 30
transmission electron microscopes (TEM), 28–30
travertine, 85
Treviranus, Ludolph Christian, 39
tumors, 83
tungsten, 30

U

unicellular organisms, defined, 22
universal joints, 38

V

vacuoles, 17
vacuum chambers, 32
vacuum systems, 31
van Leeuwenhoek, Antonie, 25, 34–36
vascular tissue, 39
Virchow, Rudolf, 7, 8, 42–43, 44, 91
viruses, 76, 81
von Waldeyer-Hartz, Heinrich Wilhelm Gottfried, 53

W

wastes, water and, 11
water, 11
Watson, James, 60, 61–65
waxes, 12
Weismann, August, 56
Wilkins, Maurice, 61–65
Wolff, Caspar Friedrich, 38–39
Wren, Christopher, 36

X

X-ray crystallography, 61

Z

Zeiss, Carl, 25–27

About the Author

Allan B. Cobb is a full-time freelance writer living in Central Texas. When not writing, he spends his time exploring nature through backpacking, caving, hiking, kayaking, and sailing. Before becoming a writer, he worked as a scientist in the environmental field. Cobb has a strong background in biology, chemistry, and geology. He has worked in the field and in the laboratory, and has aided in the management and conservation of a number of different endangered species. Currently, he is assisting on an archeological project and is a consultant in the fields of biology, chemistry, and geology. Cobb enjoys traveling, and has traveled through the United States, Canada, Mexico, Guatemala, and Belize, as well as parts of the Caribbean and Europe.